The Paper Puzzle Book
All You Need is **PAPER!**

The Paper Puzzle Book
All You Need is **PAPER!**

Ilan Garibi

David Goodman

Yossi Elran

World Scientific

NEW JERSEY · LONDON · SINGAPORE · BEIJING · SHANGHAI · HONG KONG · TAIPEI · CHENNAI · TOKYO

Published by

World Scientific Publishing Co. Pte. Ltd.

5 Toh Tuck Link, Singapore 596224

USA office: 27 Warren Street, Suite 401-402, Hackensack, NJ 07601

UK office: 57 Shelton Street, Covent Garden, London WC2H 9HE

Library of Congress Cataloging-in-Publication Data
Names: Garibi, Ilan, author.
Title: The paper puzzle book : all you need is paper! / by Ilan Garibi, David Goodman, Yossi Elran.
Description: New Jersey : World Scientific, 2017. | Includes bibliographical references.
Identifiers: LCCN 2017029413| ISBN 9789813202405 (hardcover : alk. paper) |
    ISBN 9789813202412 (pbk. : alk. paper)
Subjects: LCSH: Paper work. | Puzzles.
Classification: LCC TT870 .G269 2017 | DDC 745.54--dc23
LC record available at https://lccn.loc.gov/2017029413

British Library Cataloguing-in-Publication Data
A catalogue record for this book is available from the British Library.

Copyright © 2018 by Ilan Garibi, David Goodman & Yossi Elran

All rights reserved.

For any available supplementary material, please visit
http://www.worldscientific.com/worldscibooks/10.1142/10324#t=suppl

Printed in Singapore

Introduction

We live in an ultrafast, extreme world. Our minds are constantly bombarded with electronic information from smartphones, computers and tablets, overloading our brains with mostly useless pieces of data that spill over into the parts of the brain that were normally used for thinking. This book gives you an opportunity to rejuvenate those brain cells that are hungry for a challenge!

Take a piece of paper. Just a piece of paper. Feel it, fold it, go with the flow and enjoy the fun! So, shut down your electronic devices (unless you are using them to read this book) and challenge your brain to a wealth of intriguing puzzles.

Ilan, David, Yossi

Preface

It had to happen. The three of us converged at a single moment in time at the Weizmann Institute of Science, during a recreational math, puzzles and games conference. Ilan is a world-renowned Origami artist. He folds paper, using the principles of the ancient Japanese art to make beautiful designs and objects. His works are showcased throughout the world. David is a puzzler. He designs and crafts puzzles, mostly out of wood, and runs many puzzle workshops. He is particularly interested in puzzles that involve the ancient 'Magen David', the six-pointed star. You will find many puzzles in the book with this theme. Yossi is a recreational mathematician, interested in puzzles and the math behind them, and believes that puzzles are an excellent way to stimulate our brains, young and all. He works at the Davidson Institute of Science Education, the educational arm of the Weizmann Institute of Science in Rehovot, Israel, using recreational math to inspire kids of all ages to get into the habit of thinking creatively.

All three of us love puzzles and we are all inspired by the great writer, philosopher and puzzler, Martin Gardner. We especially love paper puzzles. They are so simple, yet so complex. That is their great appeal. There is something really special in the 'Aha!' moment when the subtlety of a beautiful solution is revealed.

The number of puzzles that require just a piece of paper and perhaps some tape and scissors, is infinite. Trying to categorize the different kinds of paper puzzles is a difficult task. Together, the three of us, each with his different angle, collected, analyzed and invented paper puzzles of all sorts and kinds. The collection of puzzles in this book is the result of our fruitful collaboration. It is an extensive collection but, by no means, exhaustive. The infinite types of puzzles and sub-categories, and the many dilemmas concerning how to even define what a 'paper' puzzle is, create an impassable obstacle. However, we have done the next best thing, and, using a nomenclature for the different kinds of puzzles, mapped out a landscape for 'paper puzzles', giving examples for each category, and complete with solutions and explanations.

We invite you to join us in our exploration of the vast field of paper puzzles and challenge you to our favorite puzzles within.

For queries, new solutions and new paper puzzles you want to share with us, please write to: thepaperpuzzlebook@gmail.com.

Ilan, David, Yossi

Thanks

This book could not have been written without the generous help of many people.

Gadi Vishne who supplied us with many solutions, as well as puzzles. Only his ability to solve some of our crazy puzzles, made them into real puzzles!

Chagay Golan who gave some of the high-level mathematical explanations, as well as solutions, that are included in the book.

Our team of testers: Judith Golan, Michal Elran, Hanani Roichman and Yael Meron. Their help made the puzzles in this book much better.

The Davidson Institute of Science Education, the educational arm of the Weizmann Institute of Science for all the hospitality and great coffee!

There are many puzzlers around the world who contributed their own puzzles to this book, or pointed us in the direction of new genres or celebrated 'oldies'. To all those from far and wide — a big thank you.

Ilan, David, Yossi

Contents

Introduction		v
Preface		vii
How to Use This Book		xix

1. Just Folding 1

1.1.	*Square to Equilateral Triangle*	1
1.2.	*Square to Largest Possible Equilateral Triangle*	2
1.3.	*Rectangle to Regular Hexagon*	2
1.4.	*Square to Regular Hexagon*	3
1.5.	*Fold an Ellipse*	3
1.6.	*Corners Converging on a Point*	4
1.7.	*An Uncreased Square from A4*	6
1.8.	*An Uncreased $1:\sqrt{2}$ Rectangle from a Square*	7
1.9.	*Quadrisecting Rectangles into Triangles*	8
1.10.	*Quadrisecting Rectangles into Rectangles*	8
1.11.	*Quadrisecting Rectangles into Rectangles — Three Folds*	9
1.12.	*Bisecting Triangles*	9
1.13.	*Bisecting Hexagons*	10
1.14.	*Folded Letter*	11
1.15.	*Generalized Folded Letter*	11
1.16.	*Inverse Folded Letter*	12

Solutions 13

1.1.	*Square to Equilateral Triangle*	13
1.2.	*Square to Largest Possible Equilateral Triangle*	14
1.3.	*Rectangle to Regular Hexagon*	16
1.4.	*Square to Regular Hexagon*	19
1.5.	*Fold an Ellipse*	21
1.6.	*Corners Converging on a Point*	25
1.7.	*An Uncreased Square from A4*	27

1.8.	An Uncreased $1:\sqrt{2}$ Rectangle from a Square	29
1.9.	Quadrisecting Rectangles into Triangles	30
1.10.	Quadrisecting Rectangles into Rectangles	32
1.11.	Quadrisecting Rectangles into Rectangles — Three Folds	36
1.12.	Bisecting Triangles	37
1.13.	Bisecting Hexagons	39
1.14.	Folded Letter	43
1.15.	Generalized Folded Letter	43
1.16.	Inverse Folded Letter	45

2. Origami Puzzles — 47

2.1.	Black and White	47
2.2.	Oversize Black and White	48
2.3.	Rectangular Black and White	49
2.4.	Stripes	50
2.5.	Checkerboards	51
2.6.	Cell Patterns in a Grid	52
2.7.	Origami Tangram	55
2.8.	Origami Windmill Base	56
2.9.	Origami Windmill Base Shapes	57
2.10.	Origami Inside Out ISO Shapes	60
2.11.	The Square Puzzle	61
2.12.	Origami Hearts and a Square	62
2.13.	Kami Alphabet	63

Solutions — 64

2.1.	Black and White	64
2.2.	Oversize Black and White	66
2.3.	Rectangular Black and White	68
2.4.	Stripes	68
2.5.	Checkerboards	72
2.6.	Cell Patterns in a Grid	73
2.7.	Origami Tangram	75
2.8.	Origami Windmill Base	76
2.9.	Origami Windmill Base Shapes	77
2.10.	Origami Inside Out ISO Shapes	78
2.11.	The Square Puzzle	81
2.12.	Origami Hearts and a Square	81
2.13.	Kami Alphabet	82

3.	**3D Folding Puzzles**		**85**
	3.1. *Seven-Squared Net*		85
	3.2. *Maximum Cube to Wrap*		86
	3.3. *A Cube from Eight Squares*		87
	3.4. *The Russian Cube*		88
	3.5. *Join the Squares*		90
	3.6. *Closed Polyhedron from a Square*		91
	3.7. *Corners to Tetrahedron*		92
	Solutions		**93**
	3.1. *Seven-Squared Net*		93
	3.2. *Maximum Cube to Wrap*		94
	3.3. *A Cube from Eight Squares*		95
	3.4. *The Russian Cube*		98
	3.5. *Join the Squares*		98
	3.6. *Closed Polyhedron from a Square*		100
	3.7. *Corners to Tetrahedron*		103
4.	**Sequence Folding**		**109**
	4.1. *1, 2, 3, 4 on a Square*		110
	4.2. *The Eight Postage Stamps*		110
	4.3. *Complico Puzzle*		111
	4.4. *The Rascals to the Prison*		112
	4.5. *Folding Frame*		113
	4.6. *Three Vertical Cuts*		114
	4.7. *The **H** Cut*		115
	4.8. *Self-Designing Tetraflexagon*		116
	Solutions		**117**
	4.1. *1, 2, 3, 4 on a Square*		117
	4.2. *The Eight Postage Stamps*		117
	4.3. *Complico Puzzle*		118
	4.4. *The Rascals to the Prison*		121
	4.5. *Folding Frame*		122
	4.6. *Three Vertical Cuts*		125
	4.7. *The **H** Cut*		129
	4.8. *Self-Designing Tetraflexagon*		130

5. Strips of Paper 133

- 5.1. *Möbius Center Cut* 134
- 5.2. *Möbius Near-Edge Cut* 134
- 5.3. *Double Möbius* 135
- 5.4. *Two Perpendicular Bands* 136
- 5.5. *Perpendicular Band and Möbius Band* 137
- 5.6. *Two Perpendicular Möbius Bands* 138
- 5.7. *Band and S Strip* 139
- 5.8. *Strip to Pentagon* 139
- 5.9. *Strip to Hexagon* 140
- 5.10. *Knotted Strip* 140
- 5.11. *Strip to Cube* 141
- 5.12. *Strips to Star of David* 141

Solutions 142

- 5.1. *Möbius Center Cut* 142
- 5.2. *Möbius Near-Edge Cut* 142
- 5.3. *Double Möbius* 143
- 5.4. *Two Perpendicular Bands* 143
- 5.5. *Perpendicular Band and Möbius Band* 144
- 5.6. *Two Perpendicular Möbius Bands* 145
- 5.7. *Band and S Strip* 145
- 5.8. *Strip to Pentagon* 146
- 5.9. *Strip to Hexagon* 146
- 5.10. *Knotted Strip* 147
- 5.11. *Strip to Cube* 148
- 5.12. *Strips to Star of David* 149

6. Flexagons 151

- 6.1. *2 × 2 Tritetraflexagon* 152
- 6.2. *2 × 2 Hexatetraflexagon* 154
- 6.3. *2 × 2 Hexatetraflexagon II* 156
- 6.4. *2 × 3 Tetratetraflexagon* 158
- 6.5. *Trihexaflexagon* 160
- 6.6. *Hexahexaflexagon* 162
- 6.7. *Flexagon Rotor* 163
- 6.8. *Tetrahedron Hexaflexagon Rotor* 166

7.	**Fold and Cut**	**169**
	7.1. *Impossible Object*	169
	7.2. *How Many Pieces?*	170
	7.3. *Star of David*	171
	7.4. *Maximum Length*	171
	7.5. *Silhouettes*	172
	7.6. *Fold-and-Cut Square*	173
	7.7. *Fold-and-Cut Star of David*	174
	7.8. *Fold-and-Cut Hollow Star of David*	174
	7.9. *Fold-and-Cut Cross*	175
	7.10. *Fold-and-Cut **A** to **Z***	175
	7.11. *Strip to a Square*	176
	7.12. *Impossible Flap*	176
	Solutions	**178**
	7.1. *Impossible Object*	178
	7.2. *How Many Pieces?*	179
	7.3. *Star of David*	180
	7.4. *Maximum Length*	182
	7.5. *Silhouettes*	183
	7.6. *Fold-and-Cut Square*	185
	7.7. *Fold-and-Cut Star of David*	186
	7.8. *Fold-and-Cut Hollow Star of David*	187
	7.9. *Fold-and-Cut Cross*	188
	7.10. *Fold-and-Cut **A** to **Z***	189
	7.11. *Strip to a Square*	191
	7.12. *Impossible Flap*	193
8.	**Just Cutting**	**195**
	8.1. *A Coin through a Hole*	195
	8.2. *Hole through a Card*	196
	8.3. *Hole through a Card with Two Cuts*	196
	8.4. *House to Square*	197
	8.5. *Cut a Fan*	198
	8.6. *Interlocking Rings*	198
	8.7. *Cube Net to a Square*	199

	Solutions	**200**
	8.1. *A Coin through a Hole*	200
	8.2. *Hole through a Card*	201
	8.3. *Hole through a Card with Two Cuts*	202
	8.4. *House to Square*	203
	8.5. *Cut a Fan*	204
	8.6. *Interlocking Rings*	205
	8.7. *Cube Net to a Square*	206
9.	**Overlapping Paper Puzzles**	**207**
	9.1. *Overlapping Sheets in a Square — Different Sizes*	207
	9.2. *Overlapping Sheets in a Square — Equal Sizes*	208
	9.3. *Kissing Sheets — All Kiss Together*	209
	9.4. *Kissing Sheets — Couple's Kiss*	210
	Solutions	**211**
	9.1. *Overlapping Sheets in a Square — Different Sizes*	211
	9.2. *Overlapping Sheets in a Square — Equal Sizes*	211
	9.3. *Kissing Sheets — All Kiss Together*	212
	9.4. *Kissing Sheets — Couple's Kiss*	212
10.	**More Fun with Paper**	**215**
	10.1. *Torpedo (Spinning Fish)*	215
	10.2. *Rotator*	216
	10.3. *Helicopters*	217
	10.4. *Origami Screecher*	219
	10.5. *Whistle*	220
	10.6. *Boomer*	221
	10.7. *Bottle Opener*	222
	10.8. *Moving Paper*	223
	10.9. *Joining Paperclips with a Dollar Bill*	225
	10.10. *The Upside-Down Dollar Bill*	226
	10.11. *Balanced Paper*	227
	10.12. *A Glass on a Sheet of Paper*	228
	Solutions	**229**
	10.11. *Balanced Paper*	229
	10.12. *A Glass on a Sheet of Paper*	229

Appendix **231**

- A.1. *Dividing a Right Angle* 231
- A.2. *The Fujimoto Approximation — Dividing a Sheet of Paper into n Equal Parts* 233
- A.3. *The 11 Nets of a Cube* 237

References **239**

How to Use This Book

This is a puzzle book, so there is no need to read it from cover to cover. A fun way we recommend is to close your eyes and choose a page at random ...

The Chapters

The puzzles in each chapter were chosen according to a common theme. As always, there are many discrepancies, since some puzzles don't fit 'in-the-box', so to speak. Sometimes, a puzzle could fit in more than one chapter. Sometimes, putting a puzzle in the right chapter could be too much of a hint. So our definitions are quite flexible. Be flexible with us!

Difficulty

We have assigned each puzzle with a level of difficulty denoted by one to four stars. One-star puzzles are simple to solve, while four-star puzzles usually require some high-level mathematics and effort before you get the solution.

Easy	★
Intermediate	★★
Hard	★★★
Very hard	★★★★

Hints

Hints are presented in a gray background box, so you can easily avoid reading them, if you prefer that.

Paper Size and Shape

For each puzzle we have an icon representing the paper needed for the puzzle. It dictates size, shape, proportion or all.

Here is the list of the types of paper we use, and their icons:

Square paper.

Rectangular or regular printer paper.

A4 printer paper — This puzzle is based on the unique proportions of the A4 paper; $1:\sqrt{2}$.

Strip — A long, narrow rectangle.

Kami — Paper in Japanese. A square (unless stated otherwise) with only one side colored. Used primarily for Origami.

Circular paper.

Cut-out paper — With the shape defined in the puzzle.

Diagrams and Origami Basic Instructions

We follow the common practice in the Origami world, using the conventional symbols, presented here:

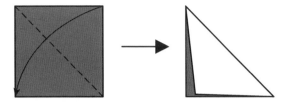

A valley fold is marked with a dashed line and a full arrow head. The paper is always folded **forward** from the tail of the arrow to its head.
Since a Kami sheet has one side colored and the other is white, this process reveals the white side.

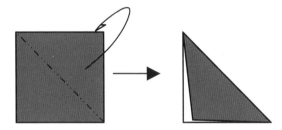

A mountain fold is marked with a dash-two dots-dash line and a hollow arrow head. The paper is always folded **backward** from the tail of the arrow to its head.

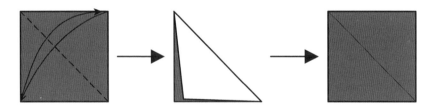

Fold and unfold are shown using two arrows, back and forth from the starting point. Note that after the paper is unfolded, a crease line appears. The new crease line is represented with a thin line.

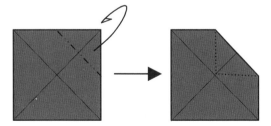

A dotted line is used to show invisible parts.

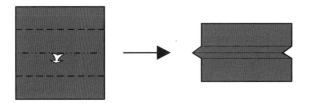

A fat arrow (hollow or full) represents a 'pull' command. The paper should be pushed inward or pulled out of the plane of the paper.

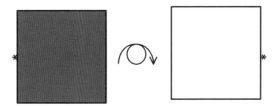

The turn over symbol is an arrow with a loop. Turn the paper over, so the left edge is now on the right (note the * position).

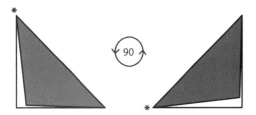

The rotate symbol shows both the amount of rotation (here by 90°) and the direction (here counterclockwise).

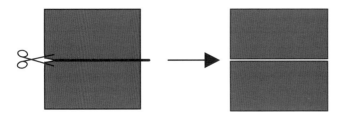

Cutting along the thick line is represented by scissors.

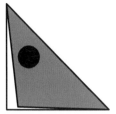

A (large) black dot represents a point to hold the model.

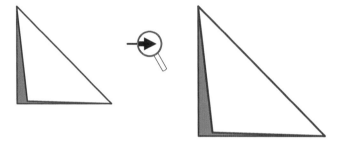

The magnifying glass symbol shows that the next image is enlarged.

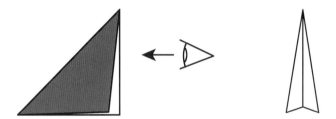

An eye and an arrow represent a change in the direction you view the model. On the left, you are asked to look at the model from the right side. The next image presents the model from that point of view.

Chapter 1

Just Folding

This chapter deals with simple folds that will unveil many puzzles. For creating the puzzles and solving them, all you need are a few folds!

Let's start with a classic that appears in the works of Henry Dudeney, one of the world's most famous puzzle makers.

1.1. *Square to Equilateral Triangle*

Level of Difficulty: ★★

Paper:

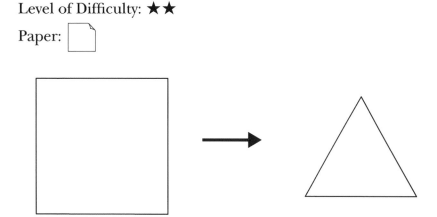

Fold an equilateral triangle from a square sheet of paper.

Since the angles in an equilateral triangle are all 60°, you have to find a way to transform the square's 90° angles into the triangle's 60° angles.

1.2. Square to Largest Possible Equilateral Triangle

Level of Difficulty: ★★★

Paper:

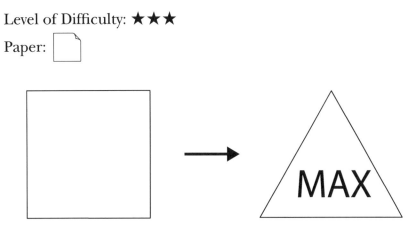

Fold the largest possible equilateral triangle from a square sheet of paper. For this puzzle, you are required to fold the largest possible equilateral triangle that can be obtained from a *given* square piece of paper. This means that the maximum amount of the square piece of paper should be used up to get the triangle.

1.3. Rectangle to Regular Hexagon

Level of Difficulty: ★★★

Paper: A4

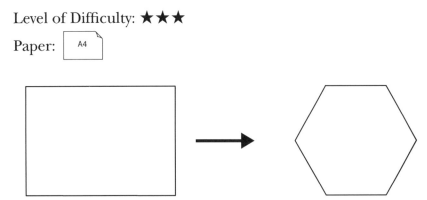

Now, it's time to try other shapes using different shapes of paper. **Fold a regular hexagon from a rectangular sheet of paper** (A4 or letter size — or any rectangle with a ratio of 1:1.1547 or bigger). A regular hexagon is one where all sides and angles are equal.

1.4. *Square to Regular Hexagon*

Level of Difficulty: ★★★

Paper:

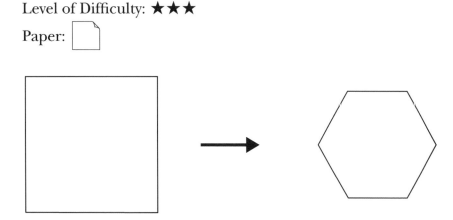

Fold a regular hexagon from a square sheet of paper. Try to maximize the hexagon's area. There are more than two solutions and one is tricky, in a way.

1.5. *Fold an Ellipse*

Level of Difficulty: ★★★

Paper:

Fold a circular piece of paper to mark out an ellipse.

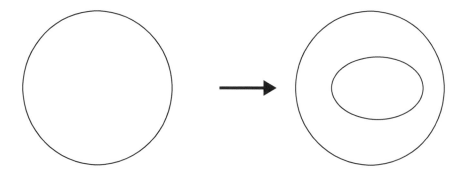

Note the subtle difference in the way this puzzle is worded. This time you are not asked to create the ellipse, only to mark it out. The mark lines should be 'drawn' by the folding creases themselves. This is a classic, mentioned by Martin Gardner in a paper dedicated to ellipses.

1.6. *Corners Converging on a Point*

Level of Difficulty: ★★★

Paper:

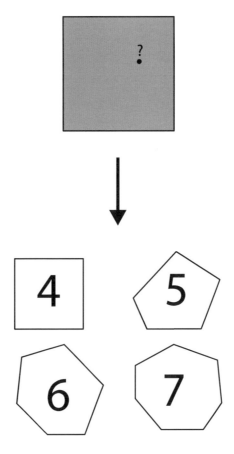

Fold the four corners of a square sheet of paper to a point, to get a:
1. square;
2. pentagon;
3. hexagon;
4. heptagon.

These shapes should be obtained by folding the corners to a point and then tucking back along the crease lines, if needed.

The solution for the square is obvious, and is presented here:

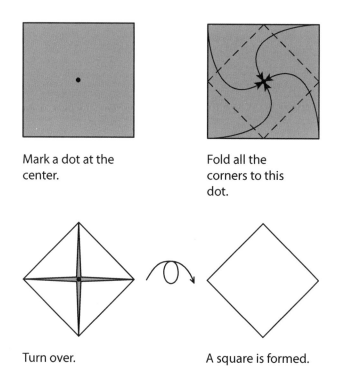

Mark a dot at the center.

Fold all the corners to this dot.

Turn over.

A square is formed.

Find the right location for the point to solve all the other polygons!

So far, we have folded different shapes from sheets of paper. Here are some puzzles where creating 'clean' uncreased shapes, is the final goal.

These *can* be cut out at the end of the folding process to give a beautiful, 'clean', uncreased shape. Only scissors are allowed. Rulers or stencils are forbidden for any purpose.

1.7. An Uncreased Square from A4

Level of Difficulty: ★★★

Paper:

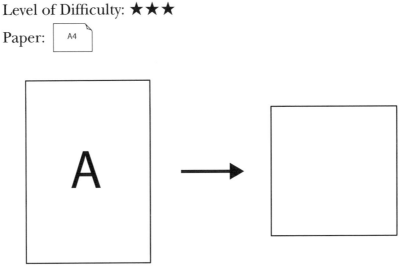

Cut an uncreased square from a single A4 sheet of paper. An uncreased square means that there are no folds, creases or bends on it. You are not allowed to use any kind of ruler, stencil or measuring apparatus. Remember, the ratio between the short and long edges of the A4 sheet of paper is $1:\sqrt{2}$.

Here is an incorrect way to solve this puzzle:

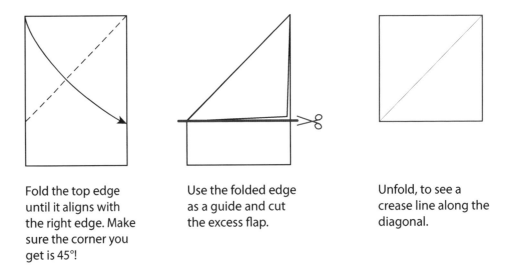

Fold the top edge until it aligns with the right edge. Make sure the corner you get is 45°!

Use the folded edge as a guide and cut the excess flap.

Unfold, to see a crease line along the diagonal.

The result is a square (which can be cut out); however, it has a fold line as its diagonal. This is not acceptable according to the puzzle rules.

So, how do we get a clean square without any creases?
There are several solutions.

> **Hint:** First, try to solve the problem starting with more than one sheet of paper. Then, try and solve starting with only one sheet of paper.

Once you've solved the original puzzle, try and find the largest possible square. One solution is based purely on geometric calculation, and another is more 'hands-on' in nature.

1.8. *An Uncreased $1:\sqrt{2}$ Rectangle from a Square*

Level of Difficulty: ★★★

Paper:

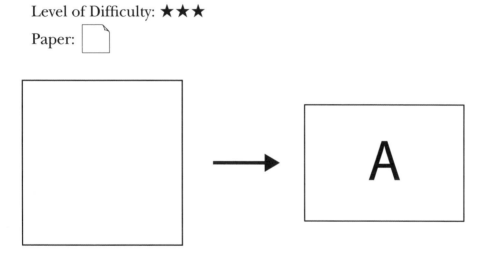

Cut an uncreased [A] proportioned rectangle from a square sheet of paper. [A] is a proportion of 1 to $\sqrt{2}$.

The real challenge is to maximize the size of the rectangle, but any solution is acceptable!

1.9. *Quadrisecting Rectangles into Triangles*

Level of Difficulty: ★★

Paper:

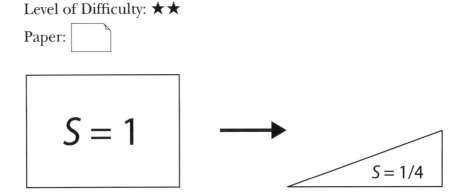

Fold a triangle from a sheet of printer paper whose area is a quarter of the area of the whole sheet, using only two fold lines. A 'pinch' made to mark a certain point on the paper is not considered a fold line. David Goodman, who independently discovered this puzzle found six different triangles.

1.10. *Quadrisecting Rectangles into Rectangles*

Level of Difficulty: ★

Paper:

Fold a rectangle from a sheet of printer paper whose area is a quarter of the area of the whole sheet, using only two fold lines. A 'pinch' made to mark a certain point on the paper is not considered a fold line. David found five different rectangles.

1.11. *Quadrisecting Rectangles into Rectangles — Three Folds*

Level of Difficulty: ★★

Paper:

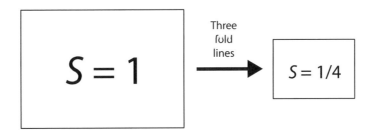

Here is a 'twist' on the previous puzzles: **find two more solutions if you are allowed to make *three* fold lines.**

1.12. *Bisecting Triangles*

Level of Difficulty: ★★

Paper:

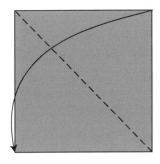
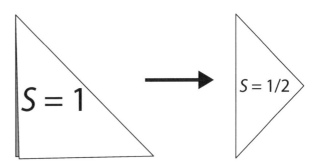

Fold a square sheet of paper along a diagonal, to form an isosceles right-angled triangle.

If we fold this triangle along the median, we get two identical isosceles right-angled triangles — each half the area of the original triangle.

Find at least one more isosceles right-angled triangle with half the area of the original triangle.

The first solution is straightforward — we can't even consider it as a puzzle. Fold the triangle in half, along its symmetry line. The area of this new triangle is half the original one.

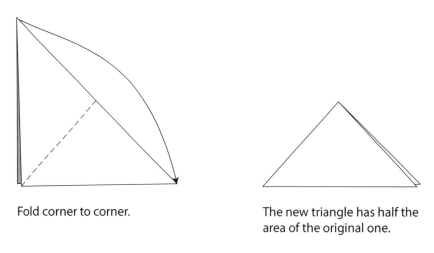

Fold corner to corner.

The new triangle has half the area of the original one.

1.13. *Bisecting Hexagons*

Level of Difficulty: ★★★

Paper:

Prepare the puzzle by cutting away a random rectangle from any rectangular sheet of paper. It must be cut away from a corner! **Find, by using folds only, a single fold line that divides the resulting hexagon into two regions of equal area.**

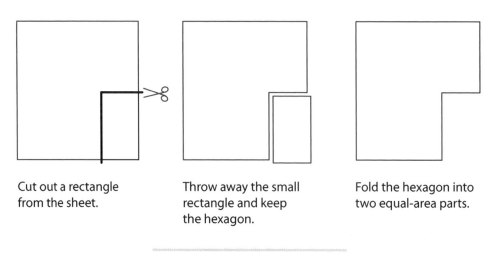

Cut out a rectangle from the sheet.

Throw away the small rectangle and keep the hexagon.

Fold the hexagon into two equal-area parts.

Scott Kim is well known for his work on ambigrams, but he also posed this famous folded letter puzzle, that Ilan Garibi later generalized.

1.14. *Folded Letter*

    Level of Difficulty: ★

    Paper:

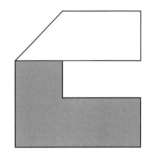

**The figure shows a cut-out letter that has been folded over once.
Which letter is it?**

L may be your first response, but there is another solution.

Hint: The folded part might be covering more of the paper than our minds want to believe. The shape itself might be oriented differently.

1.15. *Generalized Folded Letter*

    Level of Difficulty: ★★

    Paper:

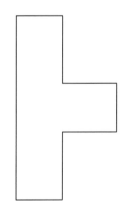

The shape of a letter is cut out from a piece of paper and then some of the paper is folded over in one or two places. The figure shows the folded shape.
What possible letters can it be?

A frequently used technique in puzzling is to ask the inverse question. In this case, the inverse problem is: given a letter of the alphabet, predict the shape that you can get if you fold the paper once (or perhaps with two folds). This isn't a very good puzzle. The number of shapes is practically infinite. A better inverse problem would be to ask for single-fold shapes that can be unfolded into a number of different letters.

1.16. *Inverse Folded Letter*

Level of Difficulty: ★

Paper:

Find the common shape from which the letters V, W, X, Y and Z can be unfolded, each using only one fold.

Chapter 1: Solutions

1.1. *Square to Equilateral Triangle*

Here's one way to do this:

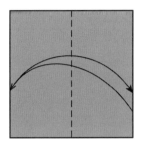

Fold edge to edge and unfold.

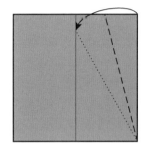

Fold the top right corner to the center line. Make sure the crease starts at the bottom right corner.

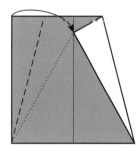

Repeat on the left.

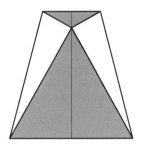

Done!

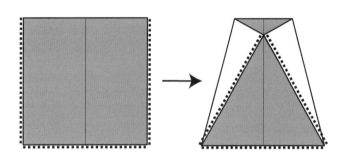

Since all the marked edges are equal, the formed triangle must have equal edges as well!

Bring the left and right top corners to converge on the center line of the square. Make sure the fold lines start at the bottom corners. The shape bounded by the folds is an equilateral triangle and the remaining paper can be folded back so they are hidden from sight. Note that since all three edges of the triangle are the original edges of the square, they are all equal by definition.

This process creates an equilateral triangle, with each angle at 60°. If we unfold the paper again and look at the angles formed by the crease at the bottom corners, the internal angle is 60° and the small exterior angle is 30°, to complete the 90° right angle. Inadvertently, we have found a way to dissect a right angle into three equal parts.

1.2. *Square to Largest Possible Equilateral Triangle*

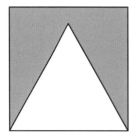
The first solution …

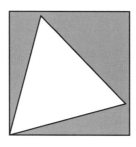
… can be rotated …

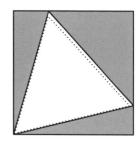
… and enlarged.

It is easy to see that the area of the white triangle in the previous puzzle is not maximized. If we rotate it along one of the common corners, we see there is room to expand the triangle, so that the other corners touch the two edges. To make the corner angle of 60°, we take advantage of symmetry. Two 15° angles have to be folded from the corner, one from each side of the required 60° angle of the triangle. These angles are folded by halving 30° angles, which we already know how to fold.

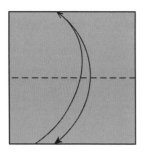
Fold edge to edge and unfold.

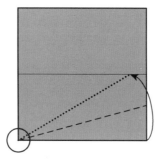
Fold the bottom right corner to the median crease line. Make sure the new fold starts at the bottom left corner!

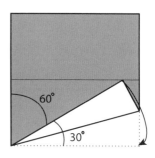
Unfold.

Chapter 1: Just Folding

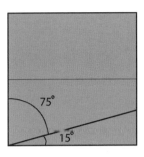

The new crease line makes a 15° angle.

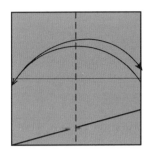

Fold right to left edge and unfold.

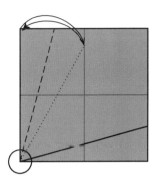

Fold top left corner to the new median crease line.
Make sure the new fold starts at the same bottom left corner as before!

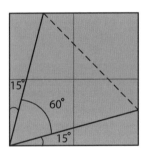

Fold a crease line that connects the two crease lines.

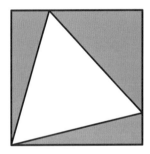

This is the largest equilateral triangle trapped in a square.

1.3. Rectangle to Regular Hexagon

This solution is based on the traditional 60° angle fold that we used before, but using another method.

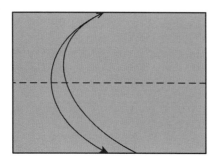

Fold in half horizontally and unfold.

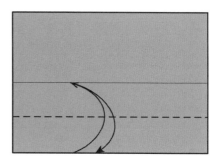

Fold the lower edge to the center line and unfold (to mark the quarter line).

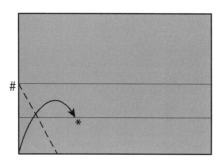

Fold the lower left corner to the quarter line (∗), and make sure the fold line starts at the (#) mark.

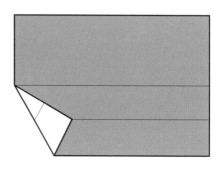

This creates a 30° angle.

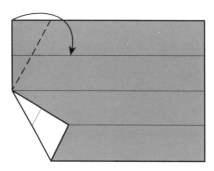

Repeat on the upper edge.

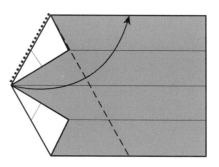

Fold the marked edge to the top edge.

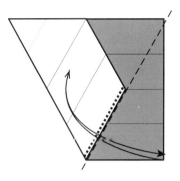

Fold the right flap along the marked edge and unfold.

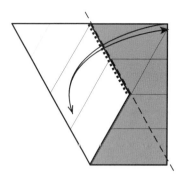

Fold the right flap along the marked edge and unfold.

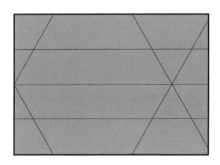

Unfold everything.

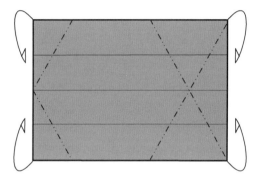

The four fold lines create a hexagon! Tuck all excess paper behind.

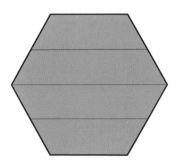

The hexagon is ready!

Take a look at these two 30°-60°-90° folded-over triangles that converge on the center line.

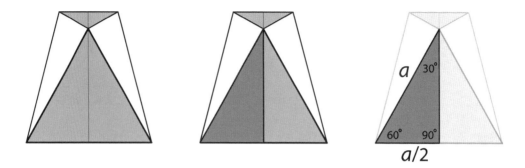

The length of the long edge of the dark triangle in the center image is the same as the length of the edge of the square, marked *a*. The length of the short edge is $a/2$ (because it meets the center line). The angles are, of course: 30°, 60° and 90°. Folding the hexagon can now be explained by using 30°-60°-90° triangles with sides: a and $a/2$; where a is the hypotenuse.

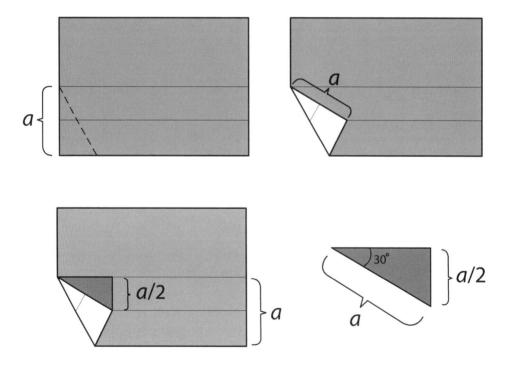

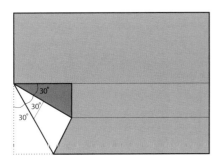 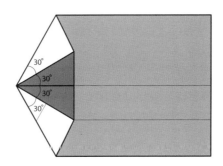

We use the same dark triangle to show the angle created by the fold is 30°. Repeat the process on the upper part to get the 120° angle needed for the regular hexagon interior angle; as all the four marked angles are 30° each.

1.4. *Square to Regular Hexagon*

Solution 1:

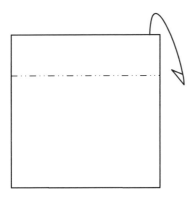 

Mountain fold the upper edge ...

... to get a rectangle.
We already know how to fold a rectangle into a hexagon!

Solution 2:

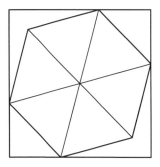

A hexagon blocked in a square.

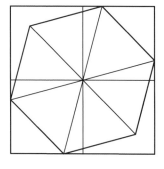

Divide it by folding edge to edge twice — bottom to top, and left to right.

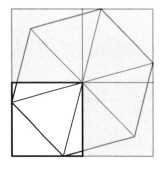

On the bottom left corner, you can see an equilateral triangle. We solved that puzzle before.

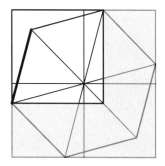

See the thick line in the marked square. The angle between the edge of the square is 15°; again, we already know how to get it. This is all you need to solve the puzzle.

In the previous puzzles, we tucked in some excess paper behind the final shape. Origamists are interested in creating a basic shape and then continue to sculpt with it. The most important part of Origami is to understand the geometry and especially the properties of different shapes so that they can be used in the folding process.

That's why we usually just 'mark out' the shape with creases in the paper so that a definitive area is bordered with the crease lines.

1.5. Fold an Ellipse

Mark a dot inside the circle and fold all points on the circumference to it.

By choosing (almost) any random point inside the circle, and folding every point on the circumference to this point, all fold lines will define an ellipse.

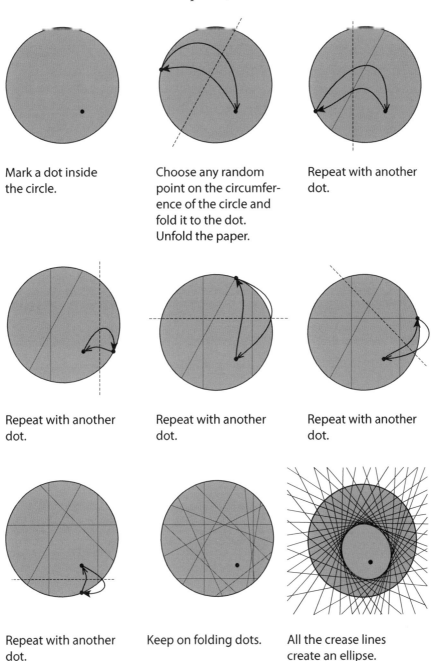

Follow-Up Activity:

If we take a more careful look at where we position the dot, we can distinguish between four cases. In each case, the final marked-out shape is different. Here are the shapes that we get for each case:

1. If the dot is positioned at a random location, we get an ellipse, as we saw above.
2. If the dot is positioned at the center, we get a circle.
3. If the dot is positioned at a random point outside of the circle, we get a hyperbola.
4. If the dot is positioned on the circumference, we get creases along all the diameters of the circle.

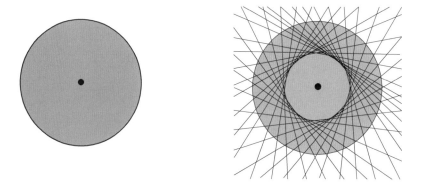

If the dot is positioned at the center of the circle, we get a circle.

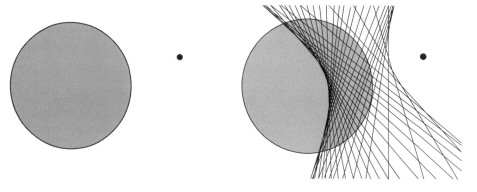

If the dot is positioned at a random point outside of the circle, we get a hyperbola.

Chapter 1: Just Folding

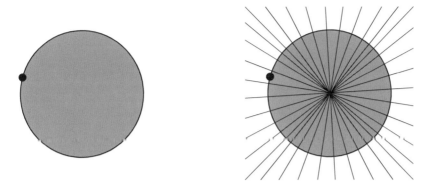

If the dot is positioned on the circumference, we get creases along all the diameters of the circle.

Follow-Up Activity:

If you fold a multitude of points on the perimeter of a square sheet of paper toward a dot inside the square, this is what you get:

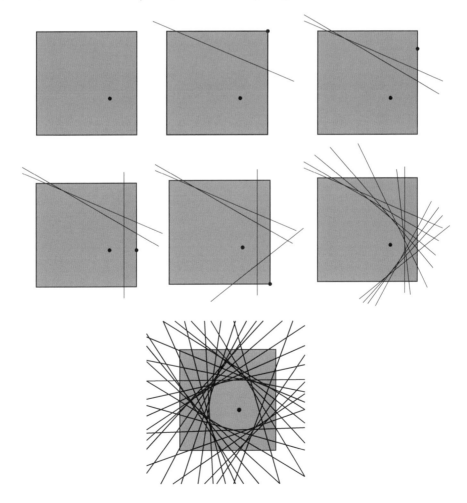

If the dot is at the exact center of the square or outside of the square, these are the shapes that we get:

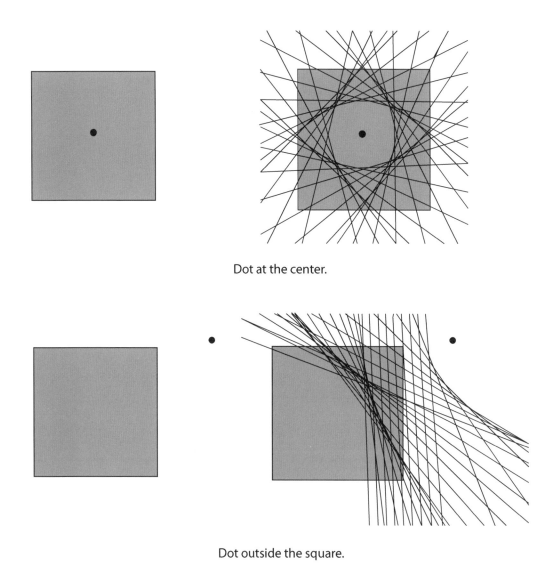

Dot at the center.

Dot outside the square.

1.6. *Corners Converging on a Point*

Another solution for the square:

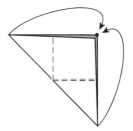

Mark a dot in the top right corner.

Fold the bottom left corner to the top right one.

Fold the other corners to the top right one.

Turn over.

A square is formed.

Here is a solution to get a pentagon:

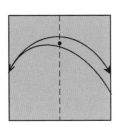
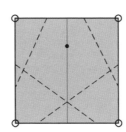
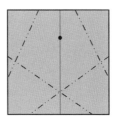

Fold edge to edge, to mark the center line. Add a dot on the upper part.

Fold a corner to the dot, and unfold. Repeat with all corners.

Now, tuck all the flaps backward, along the existing crease lines.

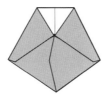

We get a pentagon. Turn over …

… to see that all corners meet at the dot.

Here is a solution to get a hexagon:

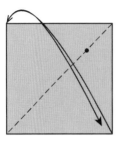 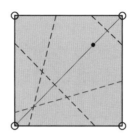 

Fold corner to corner, to mark the diagonal. Add a dot on the upper part of it.

Fold a corner to the dot, and unfold. Repeat with all corners.

Now, tuck all the flaps backward, along the existing crease lines, in the order of the numbers.

 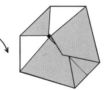

We get a hexagon. Turn over …

… to see that all corners meet at the dot.

The solution for the heptagon:

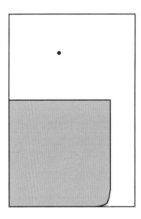 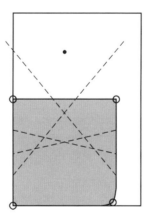 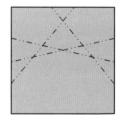

Place the square on a bigger sheet of paper, and align the bottom left corners.
Mark the dot on the bigger paper.

Fold a corner to the dot, and unfold. Repeat with all corners.

Now, tuck all the flaps backward, along the existing crease lines.

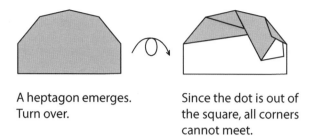

A heptagon emerges. Turn over.

Since the dot is out of the square, all corners cannot meet.

The solutions were found by Gadi Vishne.

1.7. *An Uncreased Square from A4*

Cut the paper in half and place one sheet at a right angle to the other to mark the cut line.

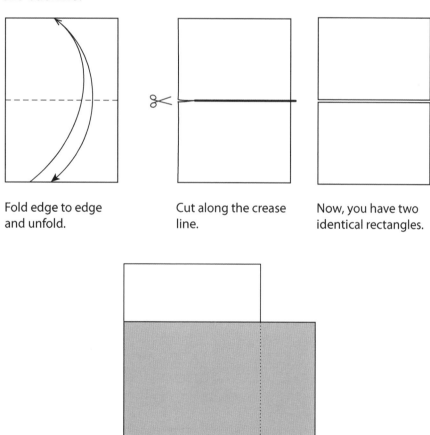

Fold edge to edge and unfold.

Cut along the crease line.

Now, you have two identical rectangles.

Now, for a harder challenge!

Cut out a square of the maximum possible size from a single A4 sheet of paper.

There are at least two solutions. One involves a calculation; the other asks for 'out-of-the-box' thinking.

Remember, the aspect ratio between the short and long edges of the A4 sheet of paper is $1:\sqrt{2}$. Inscribe inside the rectangle an isosceles triangle of edge length $\sqrt{2}$.

Note that the $\sqrt{2}$ edge of this triangle that is inside the rectangle is also the hypotenuse of the isosceles right-angled triangle of edge length 1. Hence, the top vertex of the isosceles triangle inscribed in the rectangle marks the top right corner of the square. The short side of the A4 sheet of paper is to be folded to this point.

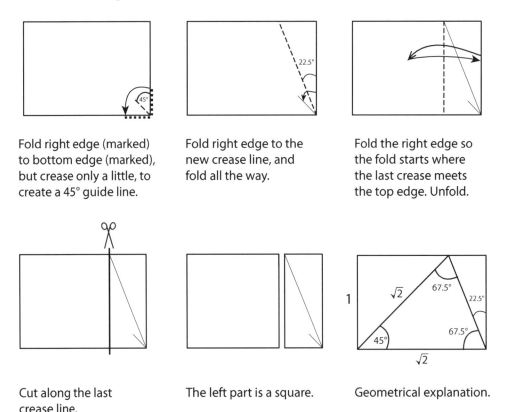

Fold right edge (marked) to bottom edge (marked), but crease only a little, to create a 45° guide line.

Fold right edge to the new crease line, and fold all the way.

Fold the right edge so the fold starts where the last crease meets the top edge. Unfold.

Cut along the last crease line.

The left part is a square.

Geometrical explanation.

This solution was found by Gadi Vishne.

An 'out-of-the-box' solution is to cut out a 'ruler' of the exact required length from the A4 sheet of paper, without damaging the area that is to be left as the smooth square as shown below.

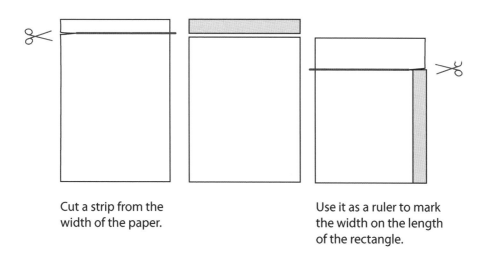

Cut a strip from the width of the paper.

Use it as a ruler to mark the width on the length of the rectangle.

1.8. An Uncreased $1:\sqrt{2}$ Rectangle from a Square

We will use the same technique as with the previous puzzle, in reverse: if we take an A4 sheet inscribed with a triangle as before, we can make an extension of the diagonal to create a square, with an edge length of $\sqrt{2}$. So we can now just reverse the procedure. Take a square sheet of paper and make two folds. One as before, at an angle of $22.5°$ (by folding the diagonal twice), and the other by folding the other diagonal ($45°$), as shown below.

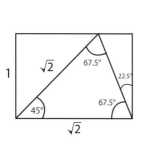 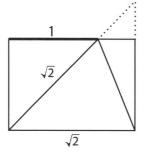 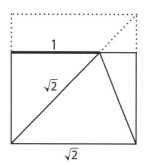

We start with the known proportion of an A4 sheet of paper.

Extend the diagonal and the right edge until their meeting point.

Complete the square by drawing the top edge and the left missing part.

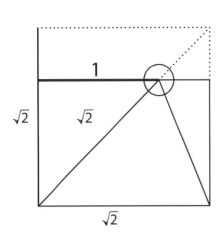
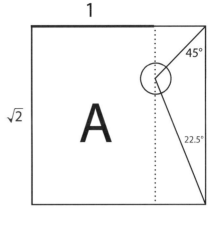

This square has an edge length of $\sqrt{2}$.
Note the marked corner.

You can find it in a square sheet by folding the 22.5° line to meet with the 45° line.
This will leave you an **[A]** proportioned sheet, uncreased, on the left!

1.9. Quadrisecting Rectangles into Triangles

There are six different triangles. The first two emerge when you fold the two diagonals:

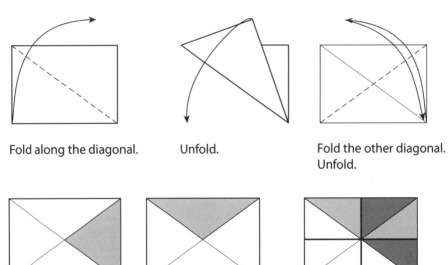

Fold along the diagonal.   Unfold.   Fold the other diagonal. Unfold.

These two triangles are equal in area, and together they are half the area of the rectangle. A simple way to show this is to divide the rectangle into four smaller ones. You can see that each triangle is divided into two triangles and all are half of the quarter rectangle.

By folding the paper in half, you get two more triangles:

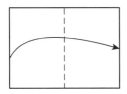
Fold edge to edge.

Fold along the diagonal of the smaller rectangle. Unfold all folds.

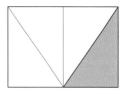
All triangles are half the area of half the rectangle.

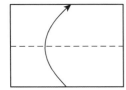
Fold edge to edge.

Fold along the diagonal of the smaller rectangle. Unfold all folds.

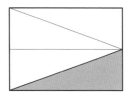
All triangles are half the area of half the rectangle.

The last two triangles are shown below:

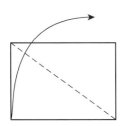
Fold along the diagonal.

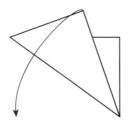
Unfold.

Solution A:

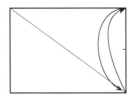
Mark with a pinch at the center of the right edge.

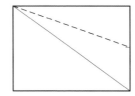
Fold along the line that connects the top left corner to the pinch mark.

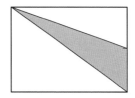
This triangle has the area of a quarter rectangle.

Solution B:

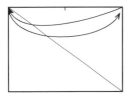
Mark with a pinch at the center of the top edge.

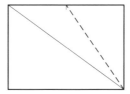
Fold along the line that connects the bottom right corner to the pinch mark.

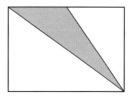
This triangle has the area of a quarter rectangle.

The explanation is based on the way you calculate the area of a triangle.

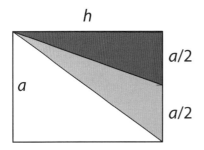
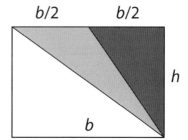

In each rectangle, both grayed triangles have the same area. For each, the base length is half of an edge (a; b) and since the height (h) is the same, the area is the same as well; according to the triangle area formula:
$$S = Base \times Height/2.$$
Dividing the grayed area into two makes it a quarter of the rectangle, because the common area is half the rectangle.

Follow-Up Activity:

In truth, there is an infinite number of solutions following this explanation. Can you find the reasoning for that?

1.10. *Quadrisecting Rectangles into Rectangles*

Getting the first three is straightforward: fold the sheet in half and then fold in half again. Since the first fold can be horizontal or vertical, three different rectangles emerge.

We get one solution by making first a horizontal fold and then a vertical fold.

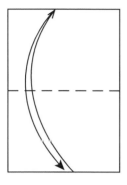
Fold bottom to top and unfold.

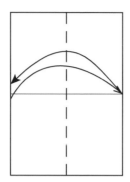
Fold left to right and unfold.

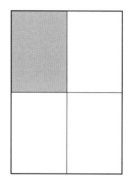
All four rectangles are equal in shape and area.

Another solution is obtained by making two consecutive vertical folds:

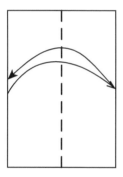
Fold left to right and unfold.

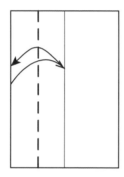
Fold left edge to the center line and unfold.

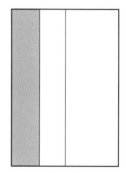
The marked rectangle is one quarter the area of the original rectangle.

The third solution with two horizontal folds:

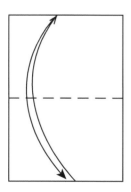
Fold bottom to top and unfold.

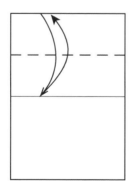
Fold top to center line and unfold.

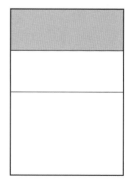
The marked rectangle is one quarter the area of the original rectangle.

More complicated solutions can also be found. Start with a pinch to mark the center of the short edge, and fold the long edge toward it. The remaining area is 3/4 of the sheet. Pinch-mark a third of this side and fold (see the appendix for marking a third). The rectangle has an area one quarter of the original square.

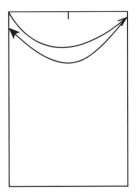

Fold left edge to the right and pinch only.

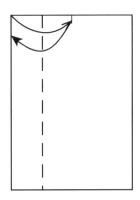

Fold left edge to the pinch mark and unfold.

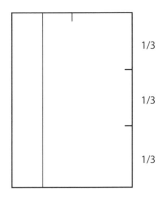

Mark the 1/3 lines on the right with pinches.

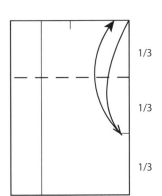

Fold the top to the farthest pinch mark and unfold.

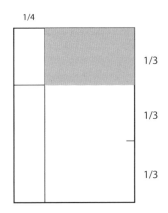

The marked rectangle is one quarter the area of the original rectangle.

If you start with the long side, you get another rectangle:

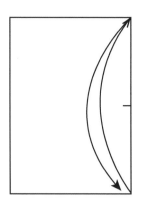
Fold bottom to top and pinch only.

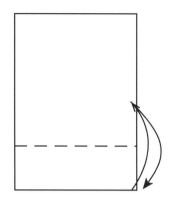
Fold the bottom to the pinch mark and unfold.

Mark the 1/3 lines on the top with pinches.

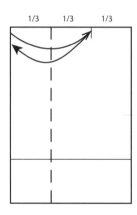
Fold left edge to the farthest pinch mark and unfold.

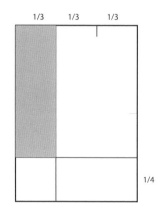
The marked rectangle is one quarter the area of the original rectangle.

1.11. *Quadrisecting Rectangles into Rectangles — Three Folds*

Starting from the last two solutions, each one of them can create another rectangle:

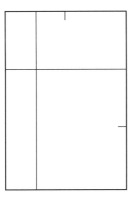 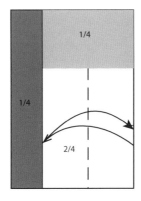 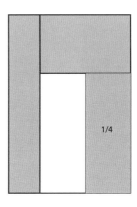

Start with the last step of the previous solution.

Fold right side to the farthest crease line and unfold.

The marked rectangle is one quarter the area of the original rectangle.

and

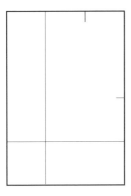 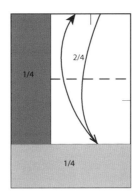 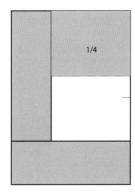

Start with the last step of the previous solution.

Fold the top to the farthest crease line and unfold.

The marked rectangle is one quarter the area of the original rectangle.

1.12. *Bisecting Triangles*

The second solution is harder to find:

Fold corner to corner and unfold.

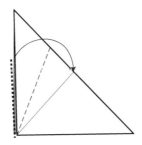

Fold the left edge to the center fold line.

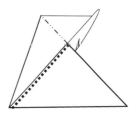

Fold the triangle flap backward.

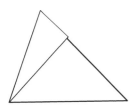

Unfold back to the first step.

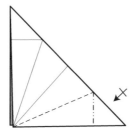

Repeat the process on the other edge.

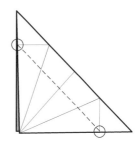

Connect the marked points with another fold.

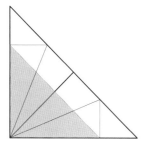
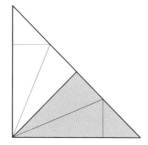
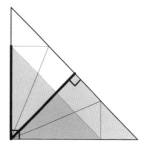

To compare the new solution (left) and the first solution (center), we merged them on the right diagram. The bold lines are identical in length. Both triangles are isosceles right-angled triangles, and have the same area.

The third solution is simpler:

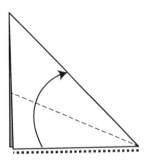 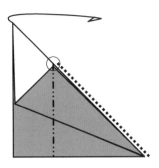 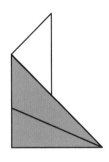

Fold the lower edge to the diagonal.

Fold backward the left edge, so the crease line starts at the marked point.

The dark triangle has the desired area.

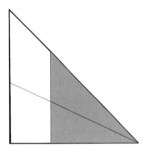 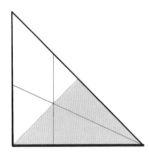 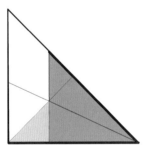

To compare the new solution (left) and the first solution (center), we merged them on the right diagram. You can see here, both triangles have the same base length and angles (45°, 90°, 45°) which means they are identical.

1.13. *Bisecting Hexagons*

The principle used here is the fact that every rectangle can be divided into two equal areas when a straight line goes through its center.

We can view this shape as if it was made out of two adjacent rectangles. Find the center of each by folding the diagonals and connect these centers with another fold line.

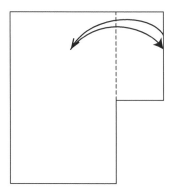

Fold and unfold the flap on the right to get two rectangles.

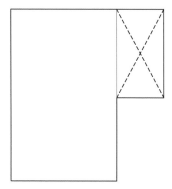

Find the center of the smaller rectangle by folding the two diagonals.

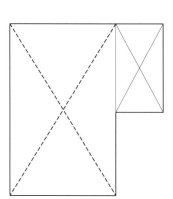

Repeat with the larger rectangle.

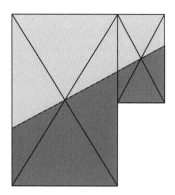

Connect the centers with a fold line.

There are other ways to divide the paper into two equal areas:

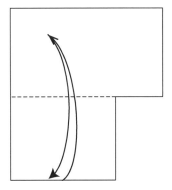

Fold and unfold the flap on the bottom to get two rectangles.

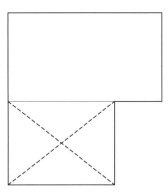

Find the center of the smaller rectangle by folding the two diagonals.

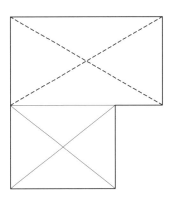

Repeat with the larger rectangle.

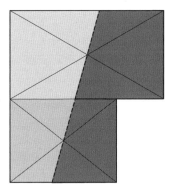

Connect the centers with a fold line.

Here is a solution that uses the rectangle that was cut out and discarded!

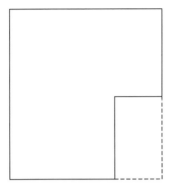

Align the cut-out rectangle with the hexagon to form the original rectangle.

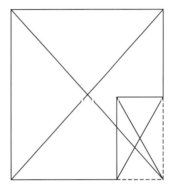

Find the center of the original rectangle, as well as the cut-out one.

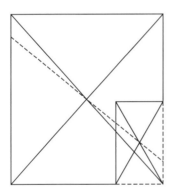

Connect the centers with a fold line.

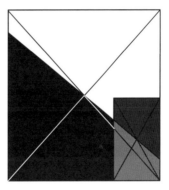

The black and white areas are the same.

Fold the original rectangle (including the realigned cut-out shape) to find its center, as before. Fold the cut-out rectangle to find its center. Once again, the line that connects the centers, divides the shape into two equal parts, even though the smaller rectangle has been cut out! This can be shown if we color the rectangles as shown in the above bottom right figure.

Subtracting the small dark gray area from the white area is equal to subtracting the small light gray area from the black area.

If we relax the condition that the shape has to be divided using only one fold line, many solutions can be found. For example, we can fold each rectangle separately as shown below. This solution was found by Roberto Gretter.

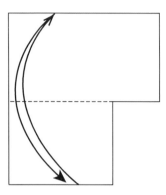

Fold bottom edge to top, and unfold.

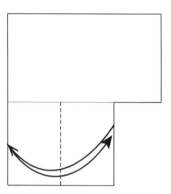

Divide the bottom rectangle into two, by folding.

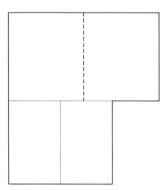

Divide the top rectangle into two, by folding.

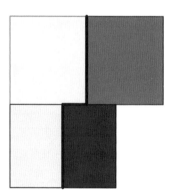

The two left rectangles are equal to the two right ones.

1.14. Folded Letter

Here is the **L**.

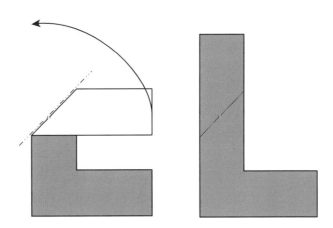

If you orient the letter upside-down, you can find the other answer — the letter **F**; which is the more challenging solution.

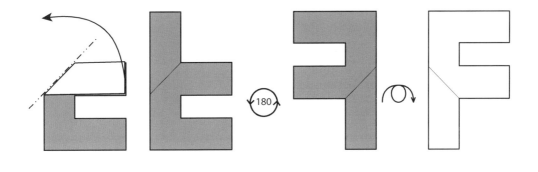

1.15. Generalized Folded Letter

At least five letters are hidden here — **H, I, F, E** and **T**.

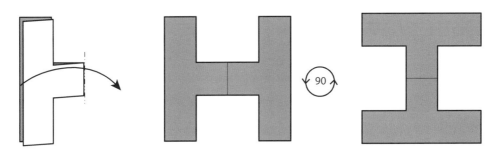

H and **I** come together.

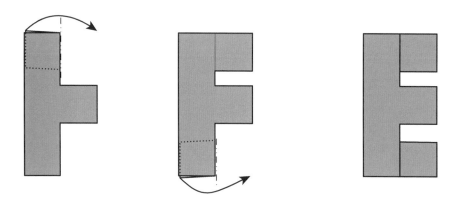

F, although a little distorted, and right after that, an **E**.

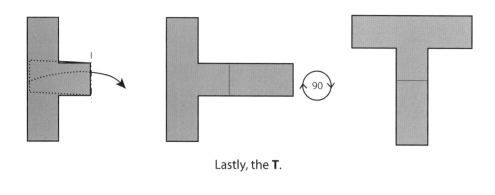

Lastly, the **T**.

1.16. *Inverse Folded Letter*

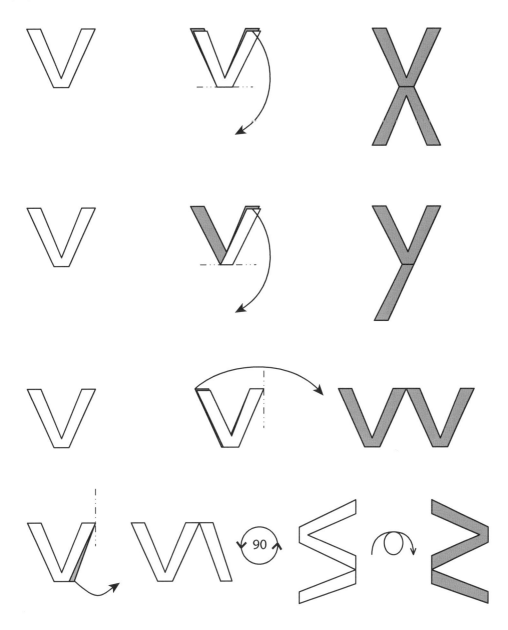

Design some more single-fold shapes that can be unfolded into a number of different letters.

Chapter 2
Origami Puzzles

Origami plays a huge part in paper manipulation. Although this is not an Origami book, we have included this chapter where we take a look at puzzles that can be solved using Origami principles and paper. The standard Origami paper, known as Kami, is colored on one side. The other side is white. Some of the puzzles exploit this change of color in the Kami paper. Others require some preparation and use Origami bases. Still others, require special-sized paper — not the usual square Kami but rectangular sheets. Mick Guy, Tom Hull, Ilan Garibi and many others combine Origami and puzzles. So, whether you are well-versed in Origami or a complete novice, we are sure you will enjoy these ...

2.1. *Black and White*

Level of Difficulty: ★

Paper:

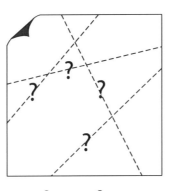

$S_{(colored)} = S_{(white)}$

Fold a sheet of Origami paper (referred to from here on as a sheet of 'Kami') and flatten it so that equal areas of colored 'black' and white parts of the paper are shown.

2.2. *Oversize Black and White*

Level of Difficulty: ★★★★

Paper:

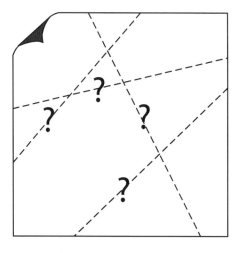

$S_{(colored)} = S_{(white)} > 1/3$

Fold a sheet of Kami, so that the visible colored part and the visible white part are both equal *and* each is larger than one third of the original size of the sheet.

2.3. *Rectangular Black and White*

Level of Difficulty: ★★★

Paper:

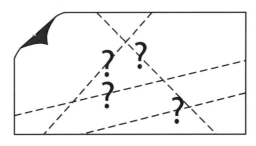

$S_{(colored)} = S_{(white)} = $ MAX %

A rectangular sheet of paper is colored on one side and white on the other. **Fold the sheet once so that both white and colored sides are visible and have an equal area.**

In the previous puzzle, we managed to get equal colored and white areas, and each was just a little more than a third of the original sheet.

What is the maximum percentage you can get in this puzzle?

2.4. Stripes

Level of Difficulty: ★★ to ★★★★

Paper:

Fold a sheet of Kami, so that three equal-area stripes are visible in sequence: colored, white and colored.

Fold a sheet of Kami, so that four equal-area stripes are visible in sequence: white, colored, white, colored.

Fold a sheet of Kami so that five equal-area stripes are visible in sequence: colored, white, colored, white, colored.

Fold a sheet of Kami so that seven equal-area stripes are visible in sequence: colored, white, colored, white, colored, white, colored.

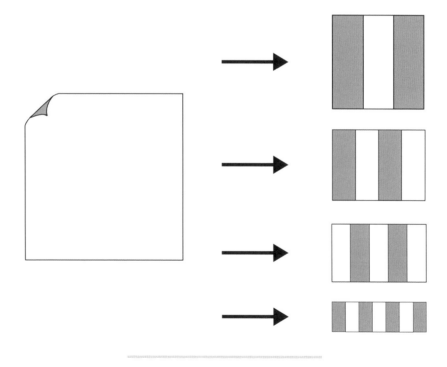

The next step up from stripes is grids. Max Hulme and Serhiy Grabarchuk, among others, pose the following puzzles.

2.5. *Checkerboards*

Level of Difficulty: ★★

Paper:

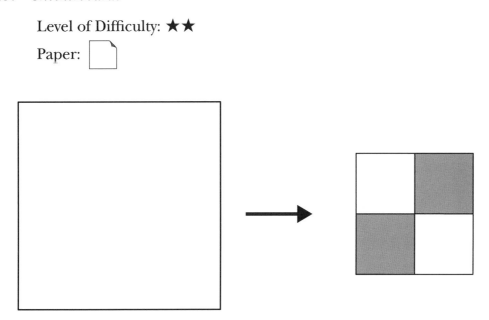

Fold a sheet of Kami into a 2 × 2 checkerboard.

As with the alternate colored stripes, one can extrapolate the checkerboard puzzle into 3 × 3, 4 × 4 and so on. This becomes a highly complicated process as you go up to the size of the 8 × 8 checkerboard, and it was first done by John Montroll (USA), in his book *Origami & Geometry: Stars, Boxes, Troublewits, Chess, & More*.

2.6. Cell Patterns in a Grid

Level of Difficulty: ★★ to ★★★★

Paper:

Fold the following patterns using a Kami sheet.

Use the minimal number of fold lines. These patterns are all the different ways to add colored square tiles onto a 3-by-3 white grid. There are 50 different patterns, not including symmetries (rotational, reflective) or negative patterns (exchanging white and black).

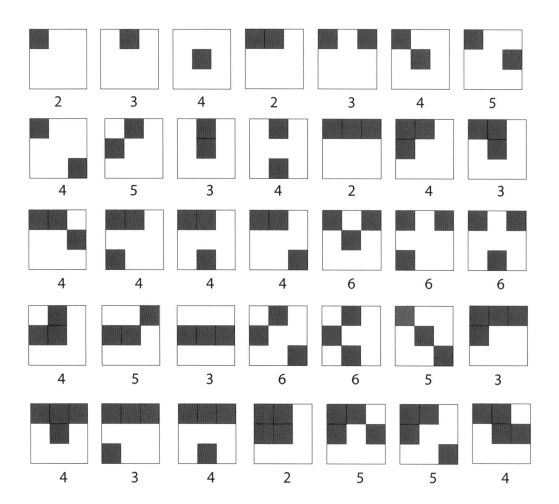

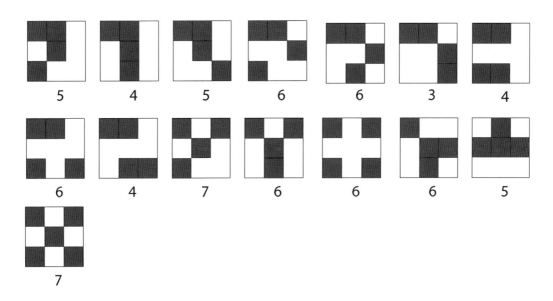

The numbers below each square show the minimum folding steps needed to solve the puzzle.

This puzzle first appeared in the International Puzzle Party (IPP) held in Luxembourg in 1996 with only the pattern of black squares (three squares along the diagonal). It was presented by Ukrainian Serhiy Grabarchuk and was expanded to the complete set of 50 (51, if you count the first one with no squares at all) challenges, by Japanese puzzlers. The solutions came from Sasaki, Kozy Kitajima and Hiroshi Yamamoto, who improved many of the solutions to the minimal number of folds. There is no known proof that these are indeed the minimal number of folds, so you can try and improve on these solutions!

Origami Tangram

Origami is the ancient art of paper folding that originated in Japan. In Origami, you usually begin with some sort of a base and then build upon that. Follow the instructions shown below, making all creases on a square sheet of paper. This creates a 12-fold square Kami which we will use in the next puzzle.

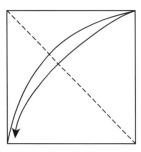

Start with the colored side down. Fold the diagonal and unfold.

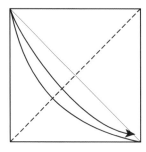

Fold the other diagonal and unfold.

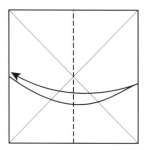

Fold left edge to the right and unfold.

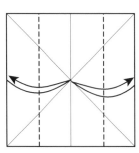

Fold both left and right edges to the center line and unfold.

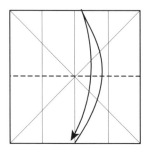

Fold bottom edge to the top and unfold.

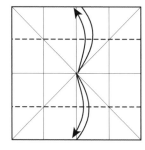

Fold both top and bottom edges to the center and unfold.

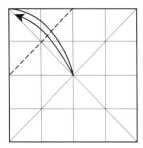

Fold one corner to the center and unfold.

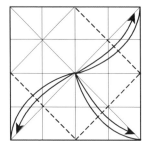

Repeat with all other three corners.

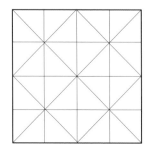

2.7. Origami Tangram

Level of Difficulty: ★

Paper:

Fold the following tangram shapes using existing fold lines only.

2.8. *Origami Windmill Base*

Level of Difficulty: ★★

Paper:

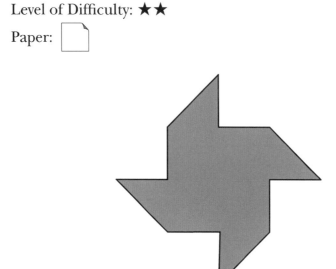

Fold the 12-fold square Kami into the Windmill Base shape as shown above.

One of the more popular bases in Origami is the 'windmill' base. It is made using only 12 folds, and we shall use it as a starting point for some puzzles.

2.9. *Origami Windmill Base Shapes*

Level of Difficulty: ★★★

Paper:

Fold the following shapes from the Windmill Base.

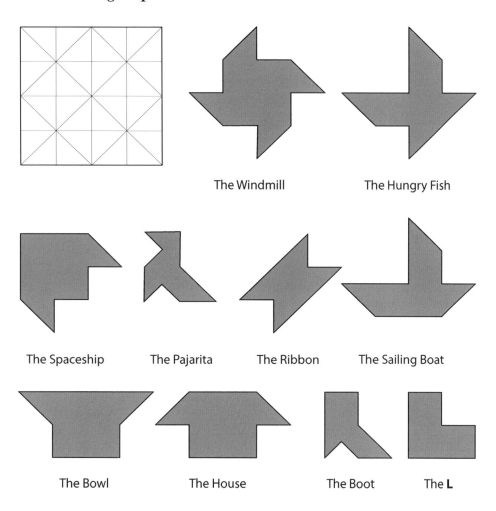

The Windmill

The Hungry Fish

The Spaceship

The Pajarita

The Ribbon

The Sailing Boat

The Bowl

The House

The Boot

The **L**

More Origami Bases

Two more popular Origami bases are the Preliminary Base and the Waterbomb Base. Interestingly, both are made out the same four creases — two valley folds along the diagonals and two mountain folds along the midsection. The only difference between the bases is the direction of the collapse — either you collect all corners together (Preliminary Base) or all midsections together (Waterbomb Base). So, if you take a Waterbomb Base and turn it inside out, you get a Preliminary Base.

Follow the diagrams to make both the bases.

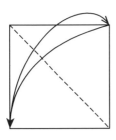

Fold corner to corner and unfold.

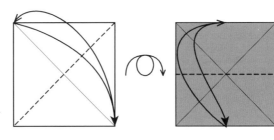

Repeat on the other corners.
Turn over.

Fold bottom to top and unfold.

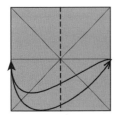

Fold left to right and unfold.

Preliminary Base

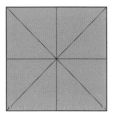

Start with the colored side up.

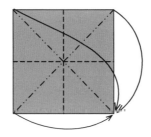

Bring all corners together to the bottom right corner.

In process.

In process.

Completed.

Waterbomb Base

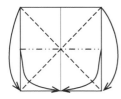

Start with the colored side down. Corners are met in pairs, while the median horizontal line collapses into the center of the bottom.

In process.

In process.

Completed.

Some Origami bases can themselves be the starting point for puzzles. For example, Ilan Garibi's Inside Out ISO Shapes and Heart and Square puzzles, and Edwin Corrie's Square puzzle that follow.

2.10. *Origami Inside Out ISO Shapes*

Level of Difficulty: ★★★★

Paper:

Fold at least one structure that has the same shape even when it is reversed (turned inside out).

A counterexample is the Preliminary Base. When reversing the shape, you do not get a Preliminary Base shape. Instead, you get a different shape — the Waterbomb Base.

Note that none of the creases changes orientation in the process — all the valleys remain as valleys, as well as the mountains! You just turn the model inside out!

From Preliminary Base to Waterbomb Base

Start with a Preliminary Base. Open it up; inside out.

In process.

Open all the way, and close in the other direction. Turn over.

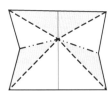

Now close it (completing the inside out movement).

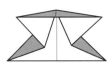

In process.

When fully reversed, you get the Waterbomb Base!

2.11. *The Square Puzzle*

Level of Difficulty: ★★

Paper:

This classic puzzle was translated to paper folding by Edwin Corrie. It is based on four identical units, to create either a square or a square within a square. Follow the instructions to create the modules.

With the four modules, create a square, and a square with a square hole inside.

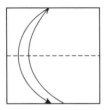
Precrease the median. Unfold.

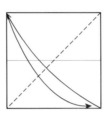
Now the diagonal, and unfold.

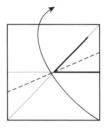
Bring the median fold line (marked) to the diagonal fold line (marked), and make sure the new fold line goes through the center.

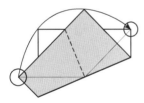
Now bring the left marked corner to the top marked corner.

Tuck the two white flaps under the top gray layer.

This is a single module. Make four.

Using the four modules, create a square; and a square with a square hole inside.

2.12. *Origami Hearts and a Square*

Level of Difficulty: ★★

Paper:

Another Origami puzzle is made from a heart module. The diagrams show how to make the module.

Fold and assemble four Origami 'heart' bases so that you can see a white square.

> **Hint:** The square can be surrounded by other shapes, as long as it is distinguishable.

Do the same, but using only two 'heart' bases.

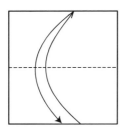
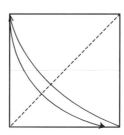
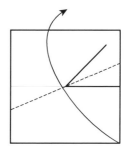

Fold edge to edge and unfold.

Now the diagonal, and unfold.

Bring the median fold line (marked) to the diagonal fold line (marked), and make sure the new fold line goes through the center.

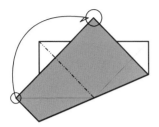 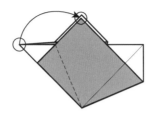 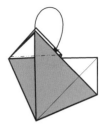

Now bring the left marked corner to the top marked corner, by folding it to the back!

Valley fold the left marked corner to the right marked corner.

Now tuck the front two layers behind.

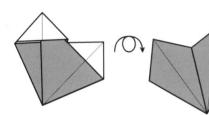

Turn it over and you get a lovely simple heart. Turn it back to get the side with the white triangles for the puzzle.
Make four units like this.

2.13. *Kami Alphabet*

Level of Difficulty: ★★★

Paper:

Using Kami sheets, try and create all the letters of the alphabet, each from a single sheet of paper. On each sheet, you may use up to four folds. The letter should be visible from the color change pattern.

This challenge was presented by Dr. Jeannine Mosely. There is a solution to this challenge for all the letters of the alphabet. In the solutions, we show four of the letters that we have found.

Chapter 2: Solutions

2.1. *Black and White*

There is more than one solution. The obvious one is to fold along one third of the paper.

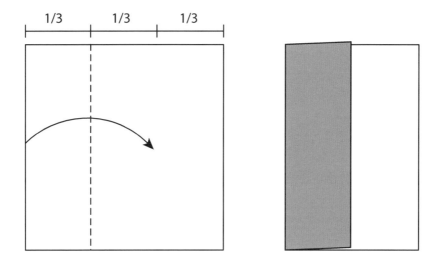

An explanation of how to find 1/3 of a given length with paper is presented in the appendix.

Here is another solution:

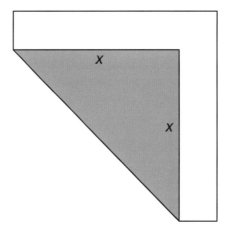

The area of the folded-over white triangle (the other side of the shaded triangle in the figure) has to be one third of the paper. This is also the area of the 'imaginary' triangle that we can trace to complete the square after folding the shaded triangle, since it is from this space where we folded in the first place. We can calculate the length of the side of the triangle according to the equation:

$$x^2/2 = \frac{1}{3}$$

where x stands for the length of the short sides of the triangle, and $\frac{1}{3}$ is the area of the triangle. $x^2 = \frac{2}{3}$, so the solution is: $x = \sqrt{\frac{2}{3}}$.

We can find the reference points for this fold using Origami only, but the process is too long for this book.

Are there more solutions?
Yes! Take a look here:

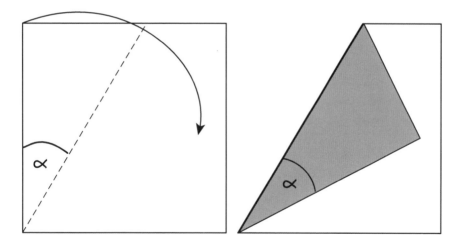

If you slowly increase the angle α, at some point the white and dark parts have to be equal. **Can you calculate this angle?**

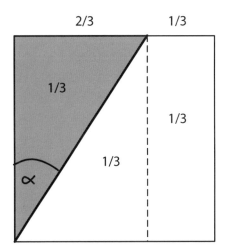

Fold the paper a third from the right. The paper is now divided into two: 1/3 on the right, and a rectangle with the area of 2/3 of the square on the left. We know a diagonal divides the rectangle into two equal parts. The area of each must be 1/3.

Using trigonometry, we can see the angle α is about 33.7°.

There are even more solutions — harder to find and prove — that we leave for you to try!

2.2. Oversize Black and White

It seems to be impossible, as we have just stated that the three parts (the visible colored part, the covered white part and the visible white part) must be equal. Still, there is a solution!

The statement above is valid only if all of the colored parts cover the white side when folding over. However, if we fold over the Kami sheet from the corner allowing the colored side to move out of the borders of the square, some of it will not cover the paper and this allows each of the parts to be larger than 1/3 of the area of the original sheet of Kami.

Here is the solution which is based on the solutions to the previous puzzle.

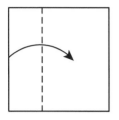
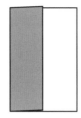

Solution 1: The 1/3 line.

Solution 2. The line creates an angle of ~33.7°.

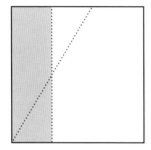

The dotted lines represent the two previous solutions. We mark the first solution area in gray.

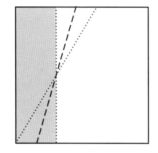

The new valley fold line creates an area on its left equal to a third of the square.

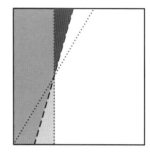

It is easy to see why — the area that is taken from the left rectangle (lower light gray triangle) must be equal to the one that is added (dark gray).

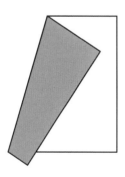

But when it is folded, it does not cover only white paper, as there is a small triangle that goes outside the square. That means the white area is bigger than 1/3!

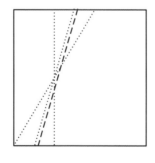

To compensate this, the valley fold line must move a little to the right and cover more of the white paper.

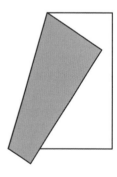

The colored area (as well as the white area) is ~0.3375, just a little larger than 0.3333 of the previous solution.

The angle is halfway between the vertical line and the slanted line. We know the angle of the latter is 33.7°, so the valley fold line should be 16.85°. Unfortunately, the location (how far to move to the right) of the line is a solution of an equation to the power of five, and cannot be reached by a folding sequence.

This calculation was done by Gadi Vishne.

2.3. Rectangular Black and White

By using a very long rectangle, you can get to almost 50% of the total area!

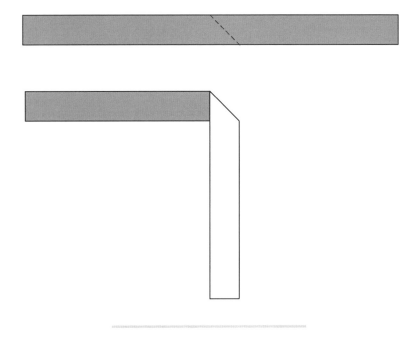

2.4. Stripes

For three stripes — you need to divide the paper into five. Folding the right and left edges inward will give the right effect.

For four stripes, you will need a more complicated folding procedure. Note that the solution doesn't need to look 'nice', as long as you get the correct color sequence.

The diagrams also show solutions for the five-stripe and seven-stripe cases.

Three stripes

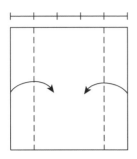 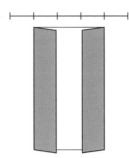

Divide the paper into five parts.
Fold the edges to the 2/5 and 3/5 pinch marks.

We get alternate colorings, three lengths wide.

Four stripes

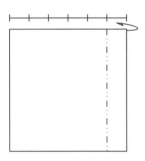 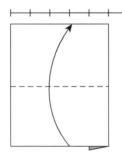 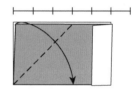

Divide the paper into six parts.
Fold the right edge backward to the second pinch mark (from the right).

Fold bottom to the top.

Fold the left corner to the bottom; only the front layer.

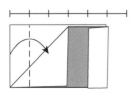 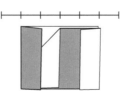

Fold the left edge to the second pinch mark (from the left).

We get alternate colorings, four lengths wide.

Five stripes

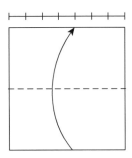

Divide into seven parts. Fold bottom to top.

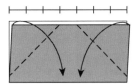

Fold both corners along the third and fourth pinch marks. The corners do not reach the bottom!

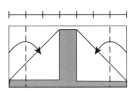

Fold both edges to the second and fifth pinch marks.

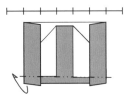

Fold the bottom backward to eliminate the colored bottom.

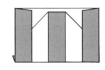

We get alternate colorings, five lengths wide.

Seven stripes

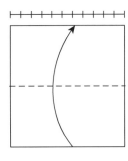

Divide into 11 parts. Fold bottom to top.

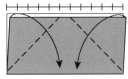

Fold both corners along the fifth and sixth pinch marks. The corners do not reach the bottom!

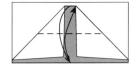

Fold the top of the front layer to the bottom and unfold.

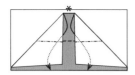

Using the new crease line as a guide, push the peak of the flap (*) inward.

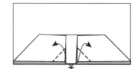

Fold the corners up, to reveal two small colored squares.

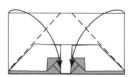

Fold the upper corners to cover those squares.

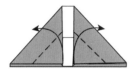

And fold the same corners up, to reveal the white part by the small squares.

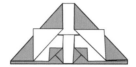

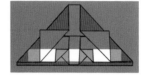

The seven-stripe area is shown in the unshaded area.

2.5. Checkerboards

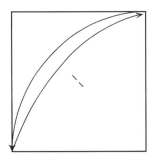

Fold corner to corner, but only pinch in the center.

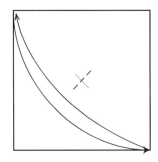

Repeat with the other corners.

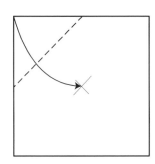

Fold a corner to the center (marked by the pinches).

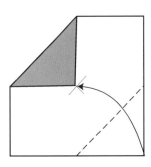

Fold the opposite corner to the center.

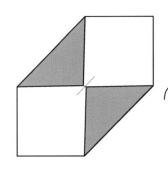

Turn over.

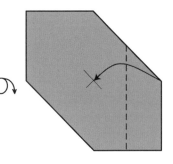

Fold the right edge to the center.

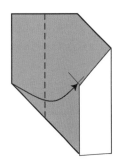

Fold the left edge to the center.

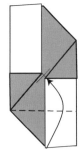

Fold the bottom to the center.

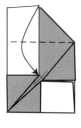

Repeat with the top.

Turn over.

The 2 × 2 checkerboard is ready!

2.6. Cell Patterns in a Grid

Here are four solutions. The rest are left as a challenge!

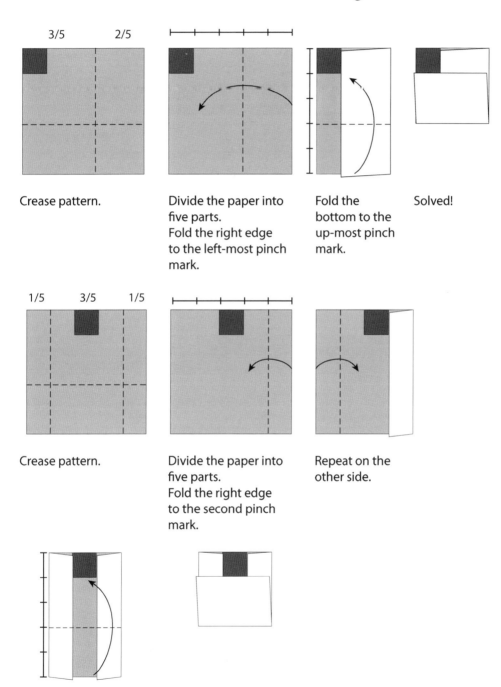

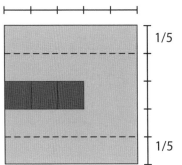

Crease pattern.

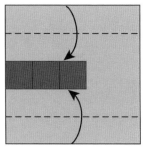

Fold the top and bottom edges to the 2/5 nearest lines.

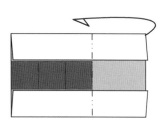

Fold the right edge backward, along the 2/5 line.

Solved!

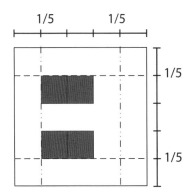

Crease pattern.

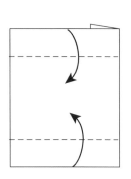

Fold the right edge backward to the 2/5 line.

Fold the top and bottom edges to the 2/5 nearest lines.

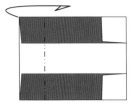

Fold the left edge backward to the 2/5 line.

Solved!

2.7. *Origami Tangram*

Here we present only two solutions.

The rest are left as a challenge for you.

2 × 3 rectangle

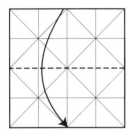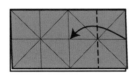

Hexagon

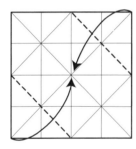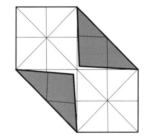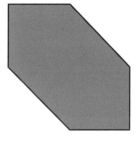

2.8. Origami Windmill Base

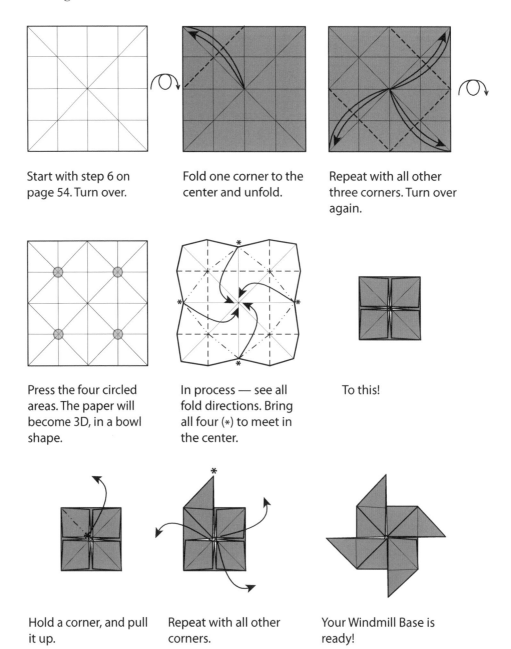

Start with step 6 on page 54. Turn over.

Fold one corner to the center and unfold.

Repeat with all other three corners. Turn over again.

Press the four circled areas. The paper will become 3D, in a bowl shape.

In process — see all fold directions. Bring all four (*) to meet in the center.

To this!

Hold a corner, and pull it up.

Repeat with all other corners.

Your Windmill Base is ready!

This shape is known in the Origami world as the Windmill Base, and as such it is a starting point for many models. One of them is the Pajarita, a Spanish bird, which is also the symbol of the Origami association of Spain.

2.9. Origami Windmill Base Shapes

The Pajarita

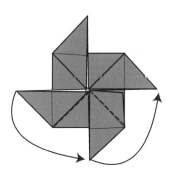
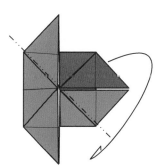
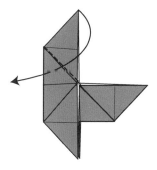

Start with the Windmill Base. Reposition the two lower flaps as shown.

Fold backward only the top right flap (marked here with darker shade).

Fold the upper flap to the left, along the valley fold line.

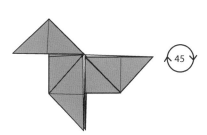
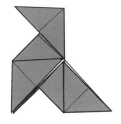

Rotate 45° clockwise.

The Pajarita silhouette is ready!

The House

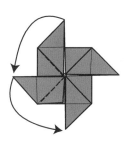
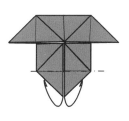

Start with the Windmill Base. Reposition the two left flaps as shown.

Fold backward the two lower flaps.

The House is ready!

The Sailing Boat

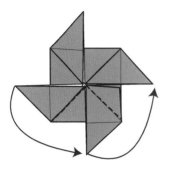 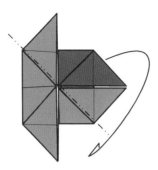 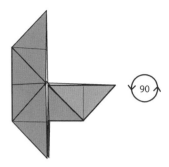

Start with the Windmill Base. Reposition the two lower flaps as shown.

Fold backward only the top right flap (marked here with darker shade).

Rotate 90° counterclockwise.

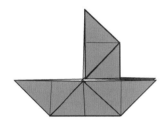 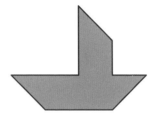

The Sailing Boat is ready!

2.10. Origami Inside Out ISO Shapes

1. Octagon solution, by Riccardo Colletto:

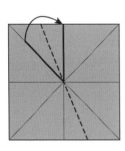 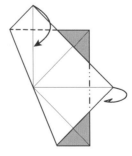 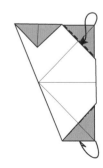

Start with a paper pre-folded as the Preliminary Base.
Fold a diagonal to the median (marked lines).

Valley fold the upper flap so the fold aligns with the original top edge. Mountain fold the right flap as well.

Repeat with the colored flaps; top as a valley (and tuck behind the top layer) and bottom with a mountain fold.

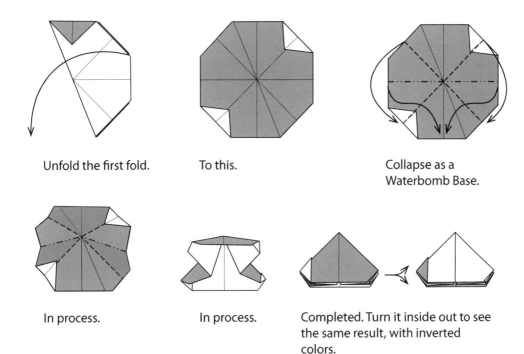

And this is an explanation why it works:

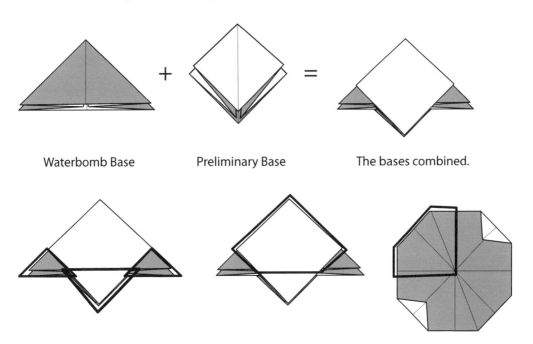

By folding all excess flaps (all the marked triangles are the same size), we get the grayed part, which creates an octagon.

2. Hexagon solution, by Dr. Saadya Sternberg:

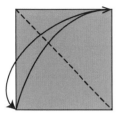

Valley fold a diagonal and unfold.

Mountain fold the other diagonal and unfold.

Fold bottom edge to top edge and pinch only.

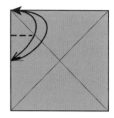

Fold top edge to the center line and pinch only.

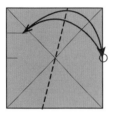

Fold right end of the central crease to the pinch mark, and make sure the new fold line goes through the center!

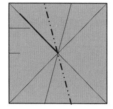

Mountain fold the diagonal (from top left to bottom right) to the last crease you made. Unfold.

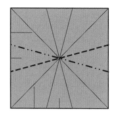

Repeat the last four steps on the vertical center line.

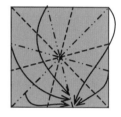

And now collapse according to the mountain-valley lines. All the mountain creases are joined!

In process.

Almost collapsed.

Ready! Turn it inside out …

… to get the same base!

2.11. *The Square Puzzle*

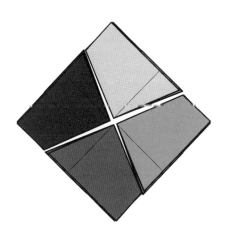
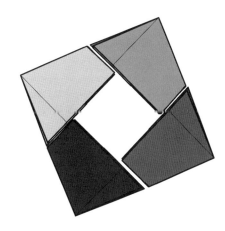

2.12. *Origami Hearts and a Square*

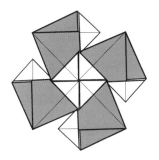
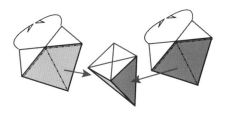

Original solution with four units

Two-unit 3D solution (by David Donahue)

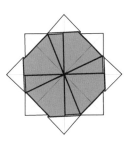

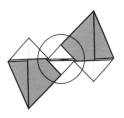

Fold the right flap into half, backward.

Make two units.

Rotate one and join the small triangles.

Two crossing squares (by David Donahue)

Two-unit solution (by David Donahue)

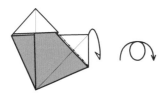

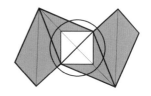

Fold the right white triangle backward. Turn over.

Make two units.

Rotate one and join the triangles.

Two-unit solution (by Bob Voelker)

2.13. *Kami Alphabet*

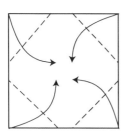
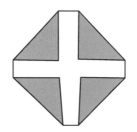
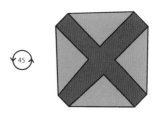

Fold all the corners almost to the center.

Rotate by 45°.

You get an **X**.

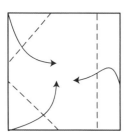
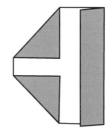
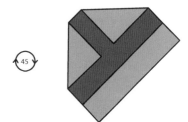

Fold the two left corners almost to the center. Fold the right edge almost to the center, too.

Rotate by 45°.

You get a **Y**.

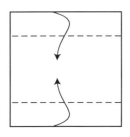 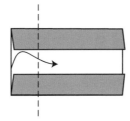 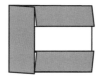

Fold top and bottom to the center, but leave a gap.

Fold the left side toward the center.

This time refer to the colored side.

You get a **C**. Rotate by 90°.

You get a **U**.

Chapter 3

3D Folding Puzzles

This chapter takes paper folding puzzles into the third dimension. Here, we will be folding cubes, tetrahedra and other polyhedra. There are quite a few well-known ways to fold a cube from a sheet of paper, especially if the sheet of paper is initially shaped as a cross. However, this wasn't always the case. The first known description of this folding appears in a German book, *Underweysung der Messung*, written by the famous 16th century architect and artist Albrecht Dürer. In his book, Dürer presented many so-called 'nets' — flat templates that can be folded into different polyhedra. Here are some fun nets and 3D puzzles for you.

3.1. *Seven-Squared Net*

    Level of Difficulty: ★★
    Paper: 1×7 ▭

There are 11 known six-squared nets that fold into a cube (see appendix). In the above, we have a seven-squared net that can still be folded into a cube of size $1 \times 1 \times 1$ (where 1 is the width of the strip). **Show how this can be done!**

You can use folds only; no cuts are allowed! Some overlaps are allowed.

3.2. *Maximum Cube to Wrap*

Level of Difficulty: ★★

Paper:

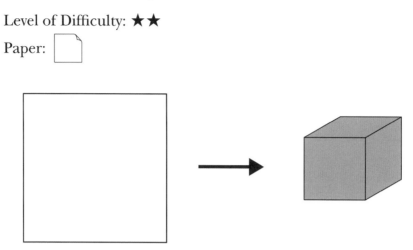

You are given a square sheet of paper. **What is the length of the side of the largest possible cube that you can wrap up using the paper?** You are not allowed to cut the paper in any way, but you can 'tuck-in' corners. Here is one solution, but there is a better one.

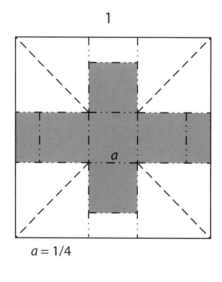

$a = 1/4$

3.3. *A Cube from Eight Squares*

Level of Difficulty: ★★

Paper:

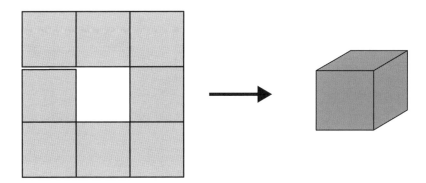

Construct a cube using the eight squares left after cutting out the center square of a 3-by-3 square sheet of paper, by folding along the lines only.

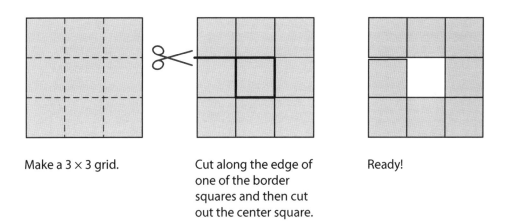

Make a 3 × 3 grid.   Cut along the edge of one of the border squares and then cut out the center square.   Ready!

If the sheet of paper is colored white on one side and blue on the other, can you control the surface color?

Can you fold a cube with all faces blue? With only one face white? Two faces white? Three faces white?

3.4. *The Russian Cube*

Level of Difficulty: ★★

Paper: 1 : 2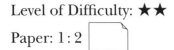

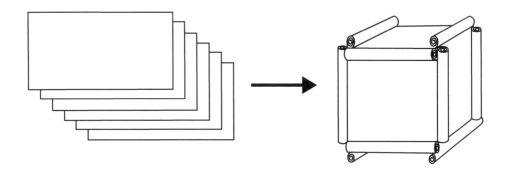

The Russian Cube is not originally made of paper. It was adopted to paper by Francesco Mancini. Use a thick paper, around 120 GSM. Make six units according to the instructions, and **try to assemble a stable cube with it.**

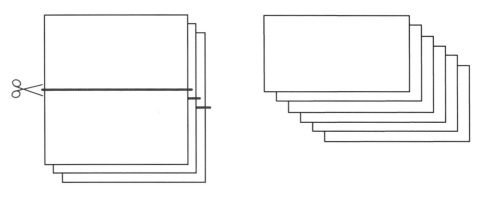

Cut three square sheets in the middle.

To make six 1 : 2 rectangles.

Chapter 3: 3D Folding Puzzles

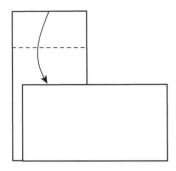
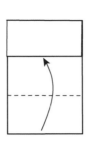
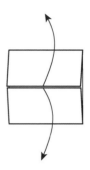

Put one unit above another as shown and valley fold the upper edge to edge of the other sheet.

Remove the other unit and valley fold the bottom to the center.

Unfold.

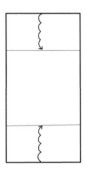
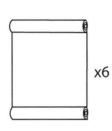

Roll both edges to the creases.

The module is ready. Make six.

Assemble the six units into a stable cube.

Other 3D Puzzles

3.5. *Join the Squares*

Level of Difficulty: ★★★

Paper:

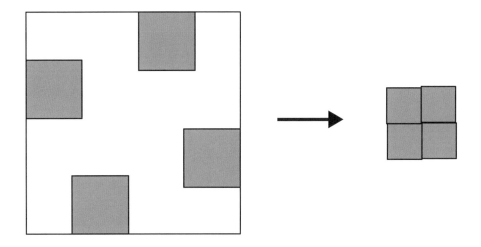

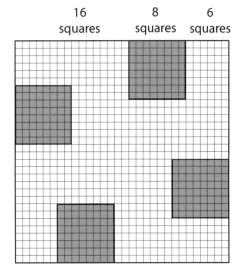

Mark a grid of 30-by-30 on a square paper. Mark four squares at a size of 8 × 8, on the edges, 6 squares from the respective edges.

Fold the square as shown and bring the four gray squares together so that a larger gray square is formed.

Hint: The area of a small gray square <u>is not</u> a quarter of the area of the large square!

3.6. *Closed Polyhedron from a Square*

Level of Difficulty: ★★ to ★★★★

Paper: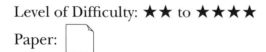

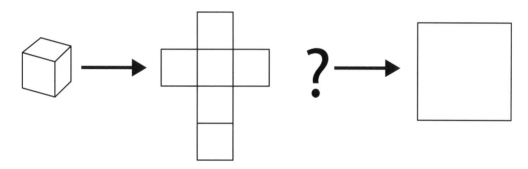

This is one of the cube's nets. Spread a cube, and you get this shape (or any other of the 11 existing shapes).

What polyhedron is spread into a square?

Fold a (closed) polyhedron from a square sheet of paper. No overlaps are allowed.

Hint: The first solution to come to mind is a pyramid.

3.7. Corners to Tetrahedron

Level of Difficulty: ★★ to ★★★★

Paper:

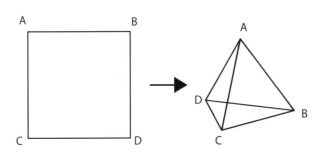

Fold a square sheet of paper so that all corners will form a tetrahedron.

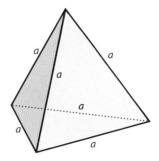

A tetrahedron is made from six equal edges. All points are at the same distance from all the others.

Fold the four corners of a square sheet of paper so that each corner is equidistant from all the others. The distance must be greater than zero.

Chapter 3: Solutions

3.1. *Seven-Squared Net*

This seven-squared arrangement is not really a valid net. A strip with seven squares cannot be folded into a six-faced cube — so, the solution should have one face made out of two overlapping squares and five free ones. Also, to get to any possible cube, we need to use some sideward folding. Two diagonal fold lines have to be made to create a position to fold a cube. Each fold along the diagonal results in a double-layered triangle. Two such triangles can be rearranged along their diagonals when folding up to form a double-layered square. To do so, they must be three squares apart. This is the shortest strip of paper you can fold into a cube!

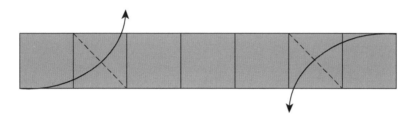

Fold the diagonals of the second and sixth squares, one flap up, the other down.

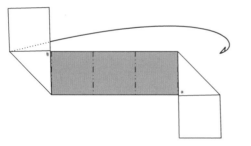

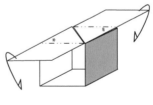

Fold the left flap backward, so the two marked (#,*) white triangles meet at the same plane.

Fold down the two flaps to close the cube.

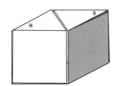

The cube is ready!

3.2. Maximum Cube to Wrap

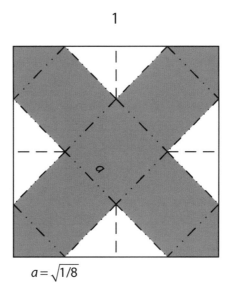

Look for a spread that cuts one or more of the six faces into small portions, so that the total area of the six squares is maximized. Here is one possible solution. Note that the four white triangles are folded inward into the cube, creasing along the line that divides each of them in half as the cube is assembled.

The side of the cube relative to the size of the side of the square sheet of paper can be calculated using Pythagoras's Theorem. It is equal to the size of the side of the square sheet of paper divided by the square root of eight. We leave it to the reader to figure out if this is the optimal solution, or if there is a better one.

3.3. A Cube from Eight Squares

With eight squares, two should overlap another two; this means you can use only two fold lines or less.

Can you overlap the relevant squares using two fold lines?

Hint: Try and create one of the 11 possible nets.

Here is the full solution. The result is a valid net for a cube, and as you can see, all faces except two will be colored.

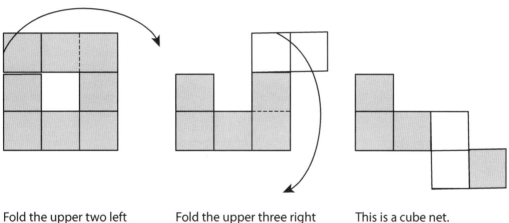

Fold the upper two left squares to the right.

Fold the upper three right squares down.

This is a cube net.

To control the number of white faces, just change the direction of the same two-fold lines. First, fold the two squares backward, so that you get a single white square in front. Fold the next fold line the same as before to get a single white face.

Here are the diagrams for all faces with the same color.

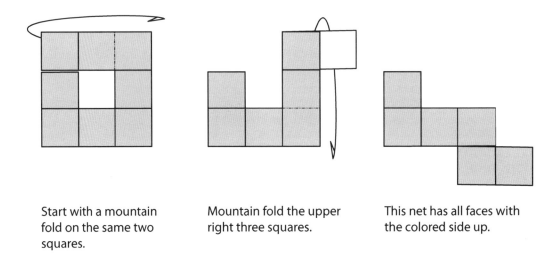

Start with a mountain fold on the same two squares.

Mountain fold the upper right three squares.

This net has all faces with the colored side up.

It is impossible to create a cube with three white sides. We leave this question unanswered as a challenge for the reader to explain why.

Follow-Up Activity:

One can cut away a square from a 3-by-3 grid in three different ways to create a missing square. Make cubes from the other two grids, following the same rules.

Solution: For a 3×3 grid, there are three types of missing square: center, side and corner.

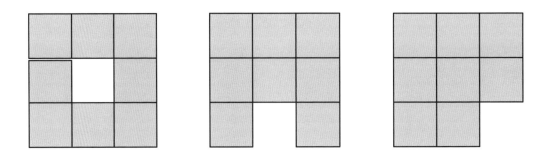

For the missing center square, there is only one valid way to cut it out, that is, by cutting through one edge of a border square.

For the missing side square, there are more. Here are some examples:

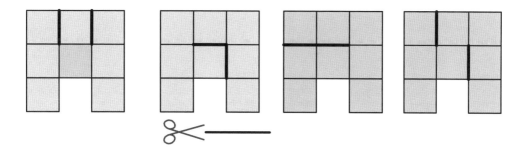

Here are more examples of nets that can be created from the possible combinations of eight squares.

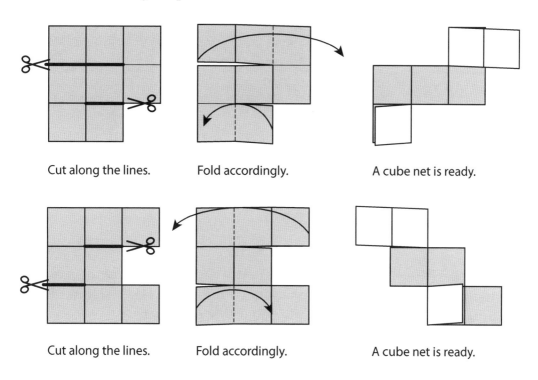

Cut along the lines.   Fold accordingly.   A cube net is ready.

Cut along the lines.   Fold accordingly.   A cube net is ready.

Try and find more ways to cut a missing square from a 3 × 3 grid and fold it into cube nets.

3.4. The Russian Cube

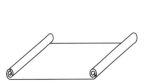

Lay one unit on the table.

Add the second as the rear face of the cube. Note where the rounded part goes.

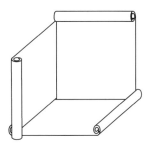

The third is the left wall.

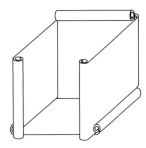

The fourth is on the right. The structure is highly unstable now.

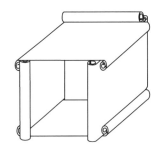

The fifth is the top. Maybe another pair of hands is useful now.

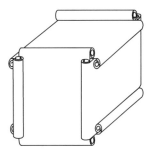

The sixth is on the front, closing the cube.

3.5. Join the Squares

The solution is based on the Origami Waterbomb Base (see diagrams in Puzzle 2.10). The fold cannot be flat, since you need to raise all four squares forward, while the other parts of the paper sink backward.

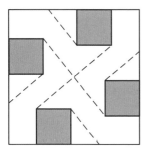

Precrease all valley folds.

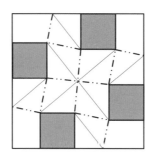

Add all mountain folds.

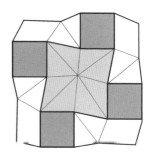

Collapse according to the fold directions (valleys and mountains). Note that the center area is actually a Waterbomb Base (in light gray).

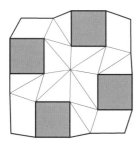

In process.

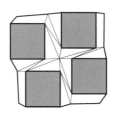

In process.

Fully collapsed.

First, make all precreases, as shown in the left-most top diagram. Then, collapse the model slowly, making sure that the inner folds create a single corner in the center of the paper, pointing downward.

3.6. Closed Polyhedron from a Square

There are endless solutions to this puzzle, but they can be divided into separate groups.

The icebreaker for this puzzle is the pyramid shown here.

A simple pyramid is folded by first bringing together two adjacent edges of the paper, and then folding in half each of the other two edges and bringing the corners to meet at the pyramid's apex. From this simple solution, it can be seen that all edges must meet the other edges, which makes a good starting point for more solutions.

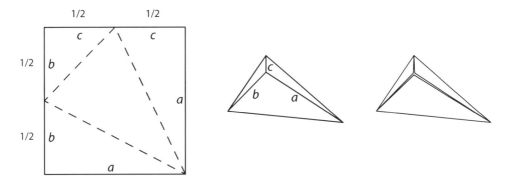

Bringing two opposite edges together creates a cylinder. Close the upper circle of the cylinder by squeezing the perimeter so that it is divided into two equal parts. Flatten the edges at the bottom, but in the orthogonal plane, that is, at an angle of $90°$. It is easiest to do this if the bottom edge is first pinched at the $1/4$ and $3/4$ marks. The resulting three-dimensional object seems peculiar. It has a linear top and bottom, perpendicular to each other, and four triangular sides. However, it is a well-known geometrical object — a tetrahedron. It only looks peculiar because we are used to a regular, equilateral tetrahedron.

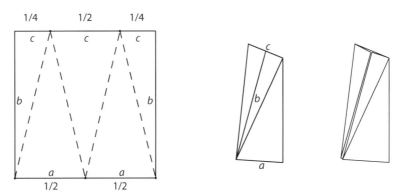

This solution can be changed slightly if you change the angle between the plane of flattening from 90° to anything but zero.

It can also be changed if you divide the paper into two parts, and do the same process on each part, as shown below in the upper solution. The result is two adjoining tetrahedra:

The bottom solution shows what happens when the center fold line is 'separated' into two, creating a gap in the center. The result is two pyramids connected through a cube.

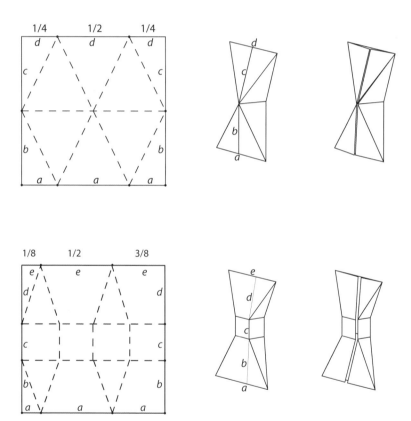

Below is the result when first folding the paper in half along the diagonal. The result is flat, but if you push the diagonal inward, you get the required third dimension in the shape of two pyramids joined.

This family of solutions is also infinite, since the angle of the pyramid corners can be changed (as long as they are smaller than 45°).

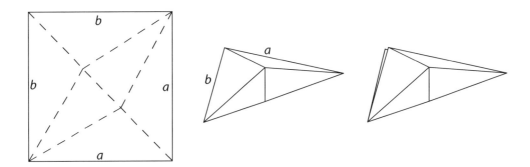

If we relax the 'polyhedron' criteria to include any three-dimensional object, there are many other families of solutions. Below is one example (directly derived from the last solution). Just curve the base, instead of using straight lines:

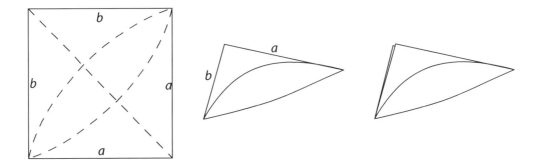

If you wonder what is the largest volume you can fold from a square sheet of paper, the solution according to Erik D. Demaine and Joseph O'Rourke is:

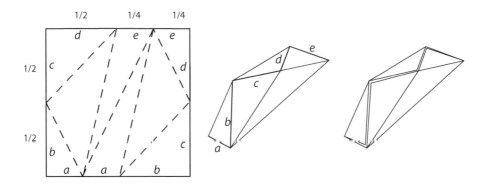

3.7. Corners to Tetrahedron

This puzzle resembles a different, well-known puzzle: **Place four points so that each is equidistant from all others.**

The solution is to assemble the points in space at the vertices of a regular tetrahedron. The third dimension *has* to be used, because there is no way to place four equidistant points on a flat plane.
Following this solution, the idea is to place the four corners of the square at the vertices of a regular tetrahedron.

There are many solutions. Here are four representative solutions.

The first solution can be divided into two.
The main idea is to fold two opposite corners in the same orientation; two above the plane of the paper, and two below.
But, where does one place the fold line, and how much do we fold?

The first solution finds the correct fold line, and asks you to fold each corner with a 90° angle.

The second solution is to fold all corners to the center; two opposite corners on the top side, and two on the bottom side. Then the question remains: how much do we need to *unfold*, so the distances of all corners from all the others are the same?

Both were solved and calculated by Hagay Golan.

The first solution is presented here:

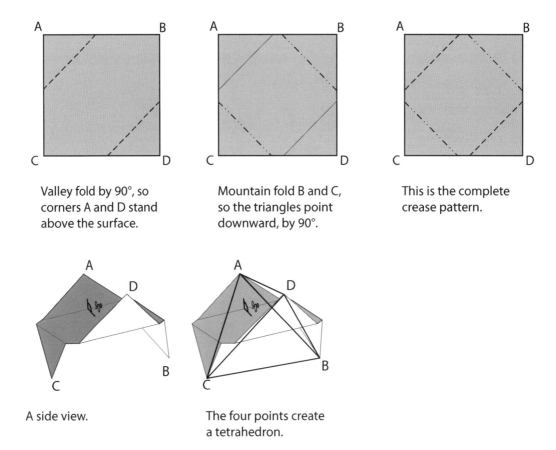

If all corners are folded by 90°, the corners form a tetrahedron.

To understand how to find the folding points, we will use the fact that a tetrahedron can be trapped in a cube. All the vertices are the extremum points of the diagonals of the cube's faces. These diagonal lines are all equal (see also the third solution, which exploits this fact).

Here is a way to find the exact location of the fold line:

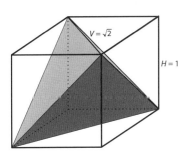

A tetrahedron can be trapped in a cube. Hence, the ratio between the height of the tetrahedron and its edge, is the same as the ratio between the height of the cube, and the diagonal of a face.

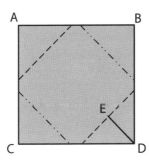

The DE line is on the diagonal. This is the distance we look for.

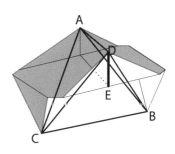

You can see that DE is actually half of *H*, the cube's height!

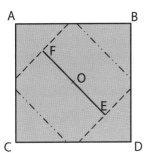

The EF line is on the diagonal.

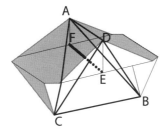

As we can see in the cube above, the EF line is actually the vertices of the tetrahedron! So EO (when O is the center) is half of *V* (the cube's diagonal).

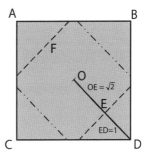

The ratio between OE (*V*/2) to ED (*H*/2) is $\sqrt{2}$ to 1!

The second solution using the same method is:

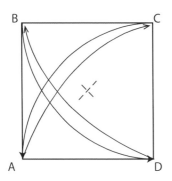

Fold corner to corner, but only pinch in the center. Repeat with the other corners.

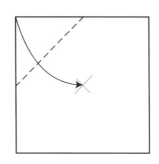

Fold a corner to the center (marked by the pinches).

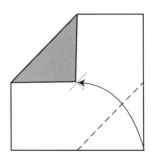

Fold the opposite corner to the center.

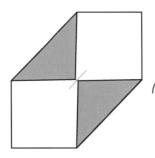

Turn over.

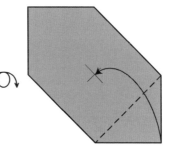

Fold bottom right corner to the center.

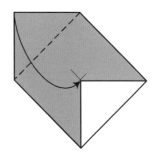

Fold top left corner to the center.

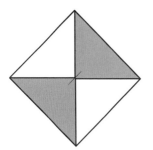

The result.

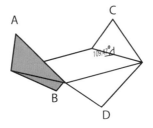

Side view.
Open all corners to a 109.47° angle.

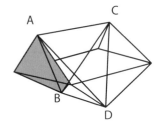

ABCD now forms a tetrahedron!

Then, slowly unfold the corners simultaneously. At some point, all distances *must* be equal. The angle is about 109.47°. The calculation process is left for the reader as an advanced challenge!

Here is another solution, based on the fact that a tetrahedron can be inscribed in a cube.

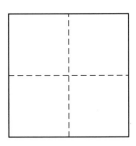

Fold the two medians and unfold.

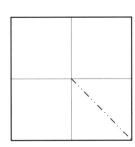

Mountain fold the diagonal of the bottom right square, and unfold.

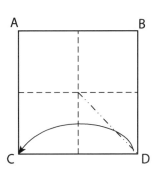

Make the model three-dimensional, by bringing point D to C; points A and B will not lay flat.

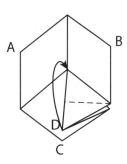

Fold point D to the center of the original square.

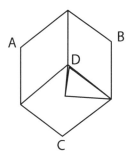

Points A, B, C and D form a tetrahedron.

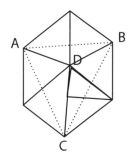

Every two corners of the square are the diagonals of a cube with an edge half the size of the original square length (AB, AC, AD, BC, BD, CD).

The last solution presents the largest tetrahedron we could find. The greatest distance we can get is at the edge of the square. By leaving all edges unfolded, the AB, BD, DC and AC keep their original distance. AD and BC are dependable (changing the distance between B and C changes the location of D), so one must calculate the accurate fold location. If AB = 1, AO is approximately 1.06. Again, we leave the calculation process to the mathematically-inclined readers!

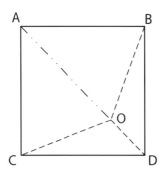

Locate O: if AB = 1, AO = 1.06. Fold all creases, and collapse the model into the desired shape.

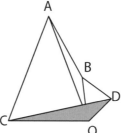
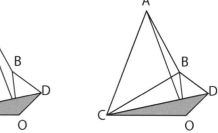

When CB is equal to AB, all corners are at the same distance from all the others.

Chapter 4
Sequence Folding

In this chapter, the objective is to fold the paper in a particular order, mainly represented by numbers or letters. You will be amazed to see how many ways you can fold, with or without cutting, the initial sheet.

Let's start with a simple example.

Take a rectangular sheet of paper, with an aspect ratio of 1:2. Fold it in half and write the number '1' on the left square and the number '2' on the right square, as shown here. The rectangular sheet of paper can be folded into two different ways: forward (bringing square '1' in front of square '2'), and backward (bringing square '2' in front of square '1'). We can define each way of folding according to the order of the numbers that appear on the squares after folding: [1, 2], and [2, 1].

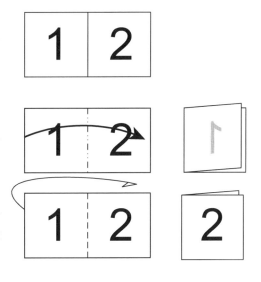

If a square sheet of paper is divided into four quarters, how many different ways are there to fold along the lines?

Since there are two fold lines, each can be folded forward or backward, you have four possible first folds and then two possible second folds — a total of eight options. Sequence folding puzzles go back at least to Henry Dudeney, and probably even before that.

4.1. *1, 2, 3, 4 on a Square*

Level of Difficulty: ★

Paper:

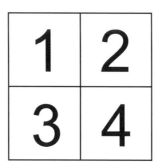

If the quarters are numbered as shown, can you prove that the configuration: [1, 2, 3, 4] is impossible?

4.2. *The Eight Postage Stamps*

Level of Difficulty: ★★

Paper: 1 : 2

The origin of this puzzle is from Henry Dudeney, a master puzzler. How many ways are there to fold this arrangement to an eight-layer thick pack the size of one small square with the top square showing the number '3'? According to Dudeney, there are 40 ways, where some require only three steps, but most require four.

Fold the rectangular sheet of paper with the number arrangement as shown, along the crease lines, so that the numbers on the final squares stacked one on top of the other are all in order; from one to eight.

4.3. *Complico Puzzle*

    Level of Difficulty: ★ to ★★★

    Paper:

'Complico' means 'to fold together' in Latin. It is a sequence puzzle on a grid of 3 × 3. There are many variations to this puzzle, with different levels of difficulty. The challenge is to fold the paper so that the squares with the letters, in sequence, spell out a word. Here are such puzzles that we found at *The Unique Projects* website.

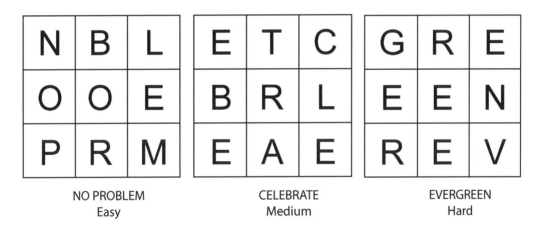

| NO PROBLEM | CELEBRATE | EVERGREEN |
| Easy | Medium | Hard |

Fold the three square sheets of paper with the letter arrangements shown as above, along the crease lines, so that the letters on the final squares stacked one on top of the other spell the words NO PROBLEM, CELEBRATE and EVERGREEN, respectively.

The level of difficulty increases with the multiple appearances of the same letter. The three E's in **CELEBRATE** add a degree of freedom we didn't have before. The hardest one we have found was **EVERGREEN**.

4.4. *The Rascals to the Prison*

Level of Difficulty: ★★

Paper:

This paper puzzle was popular during World War II. It was first published by the Vermont Publishing Company in 1942, and later popularized by Martin Gardner.

A 3 × 3 square, showed Hitler, Mussolini, and the Japanese commander Tojo in three of its squares, with jail bars in two others. The white background behind the jail bars was cut out, so you could see through them. Your task was to bring the leaders into prison. The trick was to have only two 'prison cells' available, yet three leaders worth imprisonment. So you had to choose which two they will be.

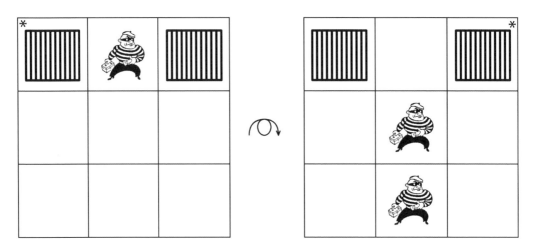

Fold the square sheet of paper along the crease lines so that the rascals become prisoners in jail. The two images shown are the back and front of the sheet.

To make this puzzle more realistic, cut out the white areas from the jail bars, so you can see the rascal behind them, when you solve the puzzle.

4.5. Folding Frame

Level of Difficulty: ★★ to ★★★

Paper: Square

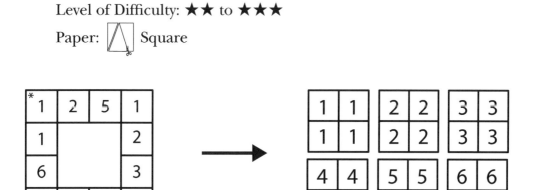

Prepare this from a 4 × 4 grid square, and remove the 2 × 2 center square. The two images shown below are the back and front of the sheet.

Fold this sheet of paper along the lines to get a 2 × 2 square with four squares, all showing the same number, for all digits from one to six.

4.6. Three Vertical Cuts

Level of Difficulty: ★★ to ★★★

Paper: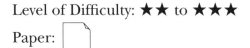

Prepare this puzzle from a 4 × 4 grid. The two images shown below are the back and front of the sheet. Add the three vertical cuts as marked. **Fold this sheet of paper along the lines to get a 2 × 2 square, that all show the same number, for all digits from one to eight.**

4.7. The *H* Cut

Level of Difficulty: ★★★

Paper:

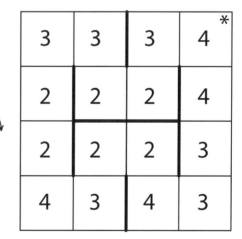

†Not in that specific order!

Another cut pattern, and more numbers to join! This time **make sure all digits from one to four appear on each side, after the paper is folded into a 2 × 2 square.**

4.8. *Self-Designing Tetraflexagon*

Level of Difficulty: ★★ to ★★★

Paper: Square

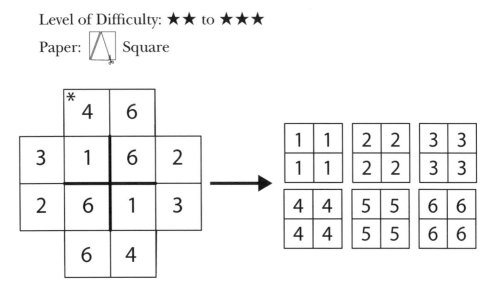

Robert E. Neale invented the following folding contraption (as far as we know), which has since become quite popular within the advertising community. Prepare this puzzle from a 4 × 4 grid square, by cutting out the four corner squares. Add the two cuts in the center. Once again, the two images below are the back and front of the sheet.

Fold this sheet of paper along the lines, to get a 2 × 2 square, all showing the same number, for all digits from one to six.

Chapter 4: Solutions

4.1. *1, 2, 3, 4 on a Square*

Since we want all numbers to be in sequence, our first fold must bring [1] and [2] together. We cannot start with the horizontal fold; since this brings [1] and [3] together. Folding the vertical fold line first brings [1] to [2], and [3] to [4]. From here, we can fold the [3, 4] couple backward or forward. The two outcomes are [1, 2, 4, 3] and [4, 3, 1, 2]. This means that the configuration [1, 2, 3, 4] cannot be folded.

4.2. *The Eight Postage Stamps*

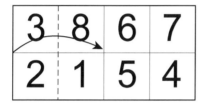

Fold left edge to the center.

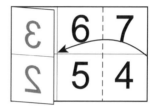

Fold right edge to the center.

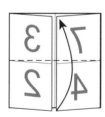

Fold the bottom edge to the top.

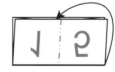

Open up the upper part a little, and bring the right lower corner between layers 3 and 8.

Finished!

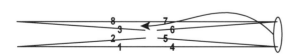

This is how the layers look from above. Insert all layers 4 to 7 in between 3 and 8.

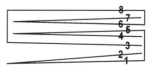

All layers are in numeric order!

Follow-Up Activity:

If you *start* with the numbers in order as shown here, you can explore all the possible options. Try to get: [1, 5, 6, 4, 8, 7, 3, 2] or [1, 3, 7, 5, 6, 8, 4, 2].

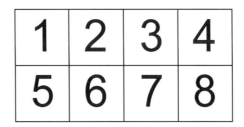

4.3. *Complico Puzzle*

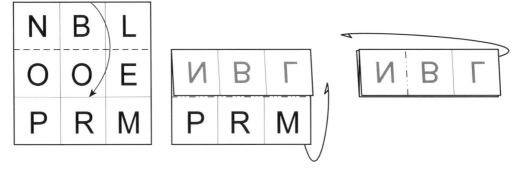

Fold upper edge [N, B, L] to the next crease line.

Fold backward the bottom edge [P, R, M].

Fold the right edge [B, L] backward along the left crease line.

Fold the left edge [M] backward along the crease line.

The letters are now in order: N, O, P, R, O, B, L, E, M.

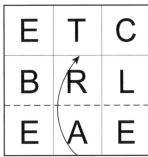 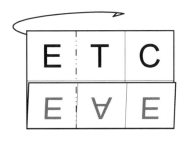 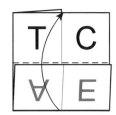

Fold bottom edge [E, A, E] to the next crease line.

Fold backward the left edge [E, E].

Fold the bottom edge [A, E] to the top.

Fold the right edge [L] to the left.

The letters are now in order: C, E, L, E, B, R, A, T, E.

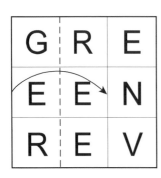
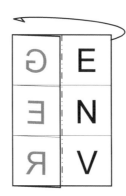
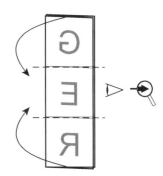

Fold left edge [G, E, R] to the next crease line.

Fold backward the right edge [E, N, V].

Bring both top and bottom flaps to a vertical position, creating a U shape. The following two steps are enlarged and the point of view is from the side.

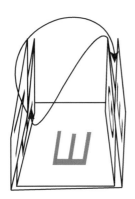
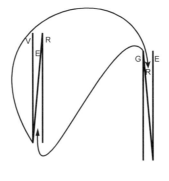

Insert the right [G] and [R] flaps between the [E] and [R] on the left, and at the same time, insert the left [V] and [E] flaps between the [E] and [R] on the right, and flatten all.

The letters are now in order:
E, V, E, R, G, R, E, E, N.

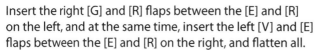

4.4. *The Rascals to the Prison*

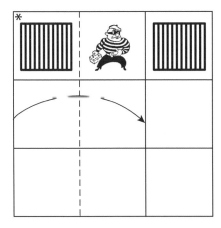

Fold left edge to the next crease line. This puts the first rascal behind bars.

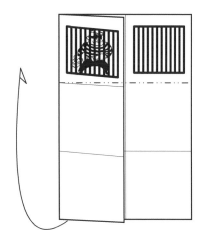

Fold the bottom backward, along the upper fold line.

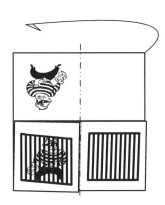

Fold the right side backward, along the center line.

Turn over.

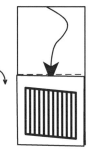

Tuck the upper part into the pocket, just one layer behind the bars, so the second one is in prison.

4.5. Folding Frame

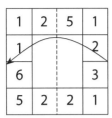

Valley fold the right edge to the left.

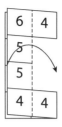

Fold the left edge back to the right (upper layer only).

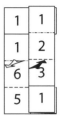

Pull the crease between [1] and [6] on the left, while pushing the crease between [2] and [3] on the right.

In process; on the left the [1] will cover the [6] and the [5]. On the right the [2] and the [3] disappear behind and the upper and lower [1]'s meet.

Solved!

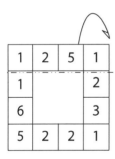

Mountain fold the top edge to the center line.

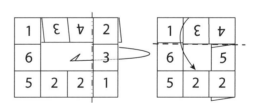

Mountain fold the right edge to the center line.

Valley fold the top edge to cover the middle row.

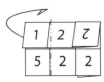

Fold backward the left edge.

Solved!

Chapter 4: Sequence Folding

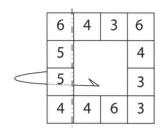
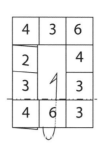
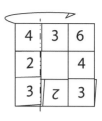

Mountain fold the left edge to the center line.

Mountain fold the bottom edge to the center line.

Mountain fold the left edge to the 2/3 line on the right.

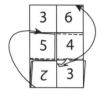

Follow the folding directions carefully! On the left, just fold the [2] on the [5]; while on the right fold the [4] on the [6] but make sure the right [3] is on top of both the [4] and [6].

Solved!

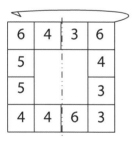
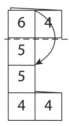

Mountain fold the right edge to the left.

Valley fold the top edge to the center.

Valley fold the top edge again to the center.

Solved!

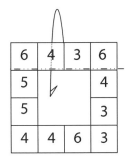

Mountain fold the top edge to the center.

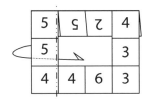

Mountain fold the left edge to the center line.

Valley fold the bottom edge to the center.

Cover the [4] with the [2] on the upper line, while valley folding the bottom [4] to cover the [2].

Solved!

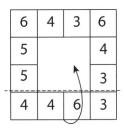

Fold bottom edge to the center line.

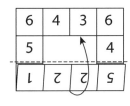

Fold again the bottom edge, **along** the center line.

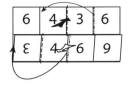

Push the crease between [4] and [3] backward on the top. Then pull the crease between [4] and [6] on the bottom. This [4] will cover the [3] on its left.

6	6
6	9

Solved!

4.6. *Three Vertical Cuts*

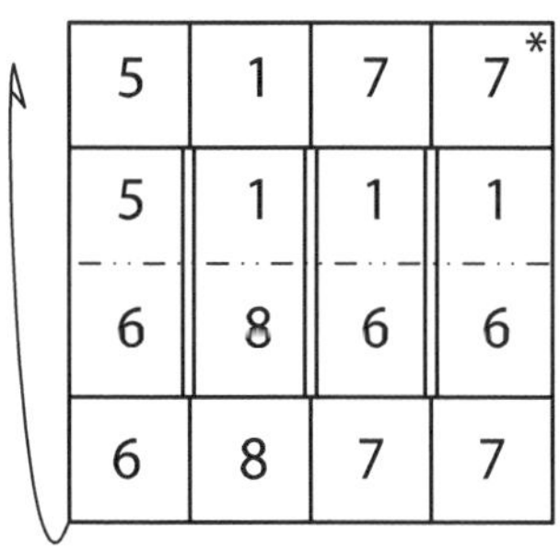

Mountain fold the bottom edge to the top.

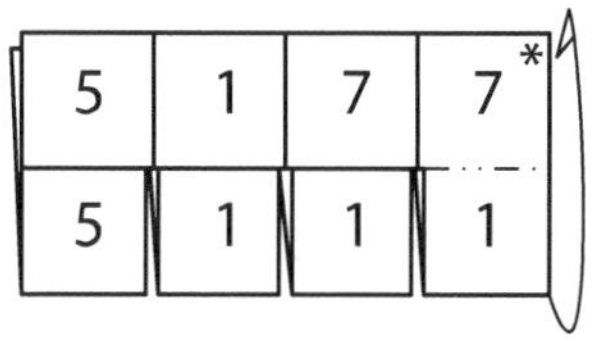

Mountain fold the right [1] flap to the top.

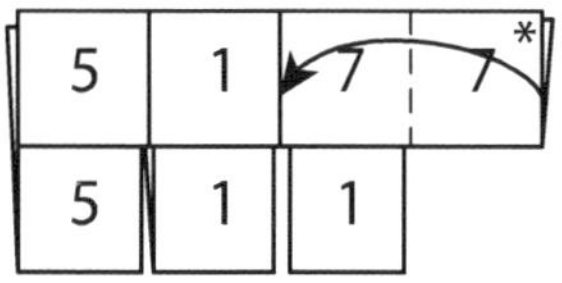

Bring this [1] forward.

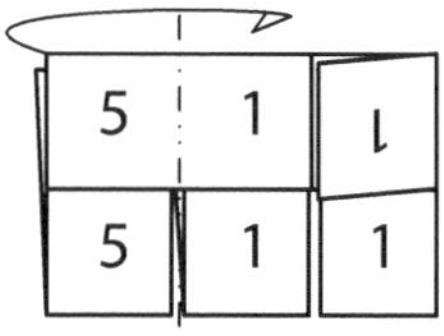

Take the [5]'s away.

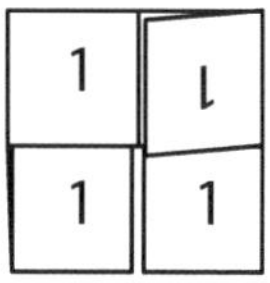

Solved!

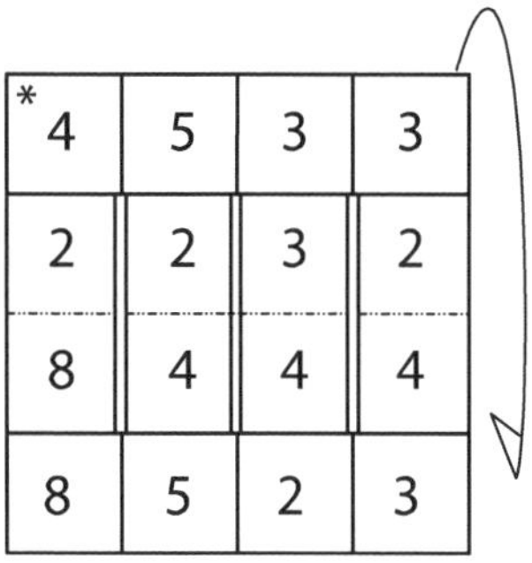

Mountain fold the top edge to the bottom.

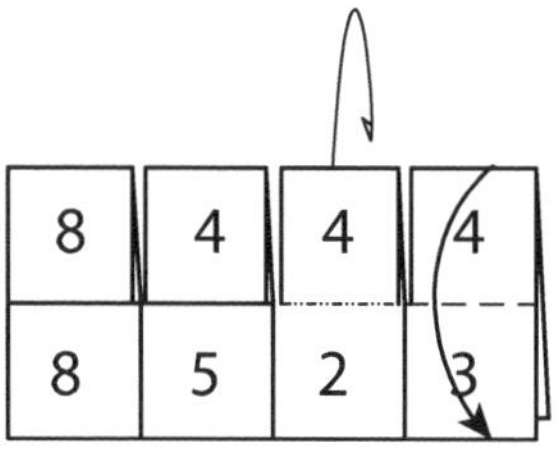

Valley fold the right-most [4], while folding backward the second [4] from the right.

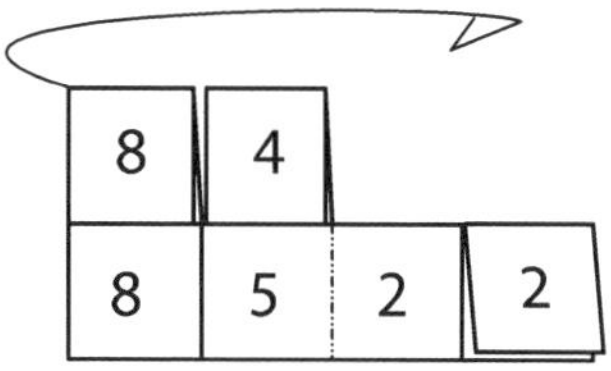

Mountain fold the left side to the right.

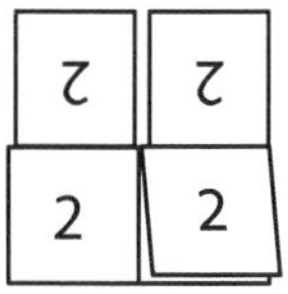

Solved!

*4	5	3	3
2	2	3	2
8	4	4	4
8	5	2	3

Mountain fold the left edge to the right.

3	3
3	2
4	4
2	3

Pull the crease between [3] and [4] on the left, while pushing the crease between [2] and [4] on the right.

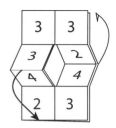

In process. Keep on to cover the [2] with the [4] on the left, and join the [3]'s on the right.

3	3
3	3

Solved!

*4	5	3	3
2	2	3	2
8	4	4	4
8	5	2	3

Mountain fold top edge to bottom.

8	4	4	4
8	5	2	3

Valley fold the left-most [8], while folding backward the left-most [4].

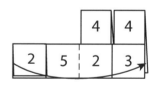

Valley fold the left side to the right.

4	4
ቱ	ቱ

Solved!

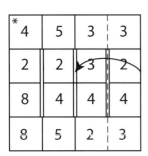

Valley fold the right edge to the center.

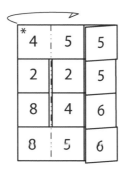

Mountain fold the left edge to the center.

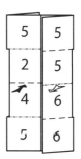

Push the crease between [2] and [4] on the left, while pulling the crease between [5] and [6] on the right.

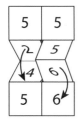

Let the [5]'s on the left meet, while the upper [6] covers the lower [6].

Solved!

Mountain fold the top edge to the bottom.

Mountain fold the second [6] from the right.

Valley fold the right edge to the center.

Valley fold the right edge to the center.

Solved!

Valley fold the bottom edge to the top.

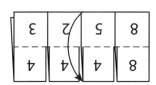

Valley fold the top edge to bottom (upper layer only).

Mountain fold the left side to the right.

Solved!

Valley fold the right edge to the center.

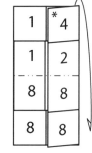

Mountain fold the left edge to the center.

Mountain fold the top edge to the bottom.

Solved!

4.7. The **H** Cut

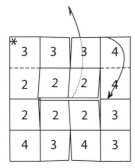

Valley fold the upper edge, while at the same time bring the upper central [2, 2] upward and backward.

Mountain fold the left edge to the center.

Valley fold the top edge to the center.

Valley fold the right side to the center.

Valley fold the lower edge to the center.

This side shows [1] to [4]. Turn over.

And here all digits appear as well. Solved!

4.8. Self-Designing Tetraflexagon

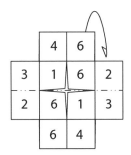
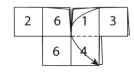
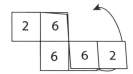

Mountain fold the top edge to the bottom.

Valley fold the [1] and [3] to the bottom.

From behind, bring the lower right flap upward to expose the [6].

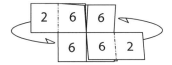

Mountain fold the [2]'s.

Solved!

Turn it over to see the [2]'s are solved, too!

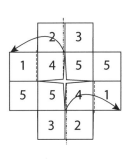
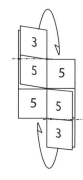
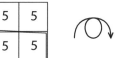

Note the folding directions! Bring the upper [4] to cover the [1] to its left, while the lower [4] covers the [1] on the right!

Mountain fold the [3]'s.

Solved!

Turn over to see the [3]'s solved as well!

Chapter 4: Sequence Folding

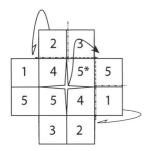

Fold [2] and [1] backward, to a 90° angle. This will cause [5*] to flip over!

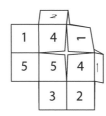

The result. Turn over.

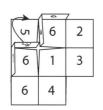

Bring the [3*] to lie down on the [5]. The model is not flat yet!

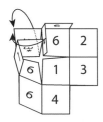

Now you have to make two steps in one move. First cover the [3*] with the [3] that is just to the right of it. While doing so, the right bottom corner of the [5] will flip (up) again to the top left corner; bringing forward [1].

The result. Turn over.

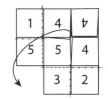

Repeat the same process on the other corner.

All [4]'s are here. Turn over.

You may get the upside-down [2] instead of a [1]. Just tuck the [2] under the [1] that is waiting below.

And you get all [1]'s, as well!

Chapter 5

Strips of Paper

At a certain point, a rectangular sheet of paper changes its name from a 'sheet' to a 'strip'. Interestingly, strips invite a whole, new field of puzzles, activities and wonders.

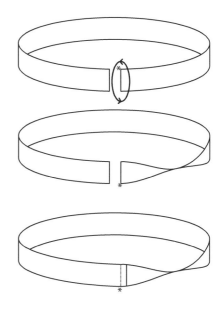

The most famous one is the Möbius strip or — perhaps more accurately called — the Möbius band. It was discovered and studied independently by two mathematicians in the 18th century: Johann Benedict Listing and August Ferdinand Möbius, after whom this surface is named.

The Möbius band is made with a strip of paper; we recommend to use an aspect ratio of approximately 1 : 20. Bring the short ends together, as if to close the strip into a ring, but just before joining them, flip or half-twist one of the ends 180°. Glue the ends together and you have a Möbius band. The Möbius band has only one side and one edge. Imagine a beetle starting to walk on the band along the center line — it walks continuously without stopping until returning to the starting point. This is a manifest of the fact that by rotating the strip's end, we merge the inner (to be) side with the outer side, making a single continuous surface.

Although these are not puzzles in the true sense of the word, we arranged this chapter in a way that keeps the surprise factor. We challenge you to try and imagine the results of the actions (mostly cutting) before you look for the solutions in the book. Clifford Pickover's excellent book, *The Möbius Strip*, is an exemplary starting point for those who want to learn more about the subject.

5.1. Möbius Center Cut

Level of Difficulty: ★
Paper: ▭

What topological object do you get when you cut a Möbius band along its center line?

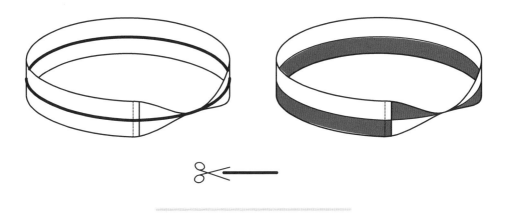

5.2. Möbius Near-Edge Cut

Level of Difficulty: ★
Paper: ▭

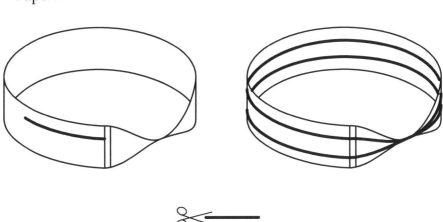

By cutting a band in the center, we ensure that the cut line returns to the starting point after a full cycle. **What happens if you do not start from the center, but at a point that is one third of the distance from the edge of the band?**

5.3. Double Möbius

Level of Difficulty: ★
Paper: ▭

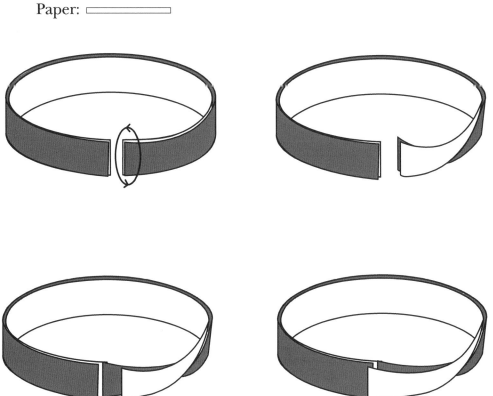

Stack two strips one behind the other, half-twist once and join the ends; gray to white, and white to gray.

What do you get?

Cutting Möbius bands can be generalized to the case of more than one 'ring'.

5.4. *Two Perpendicular Bands*

Level of Difficulty: ★
Paper: ▭

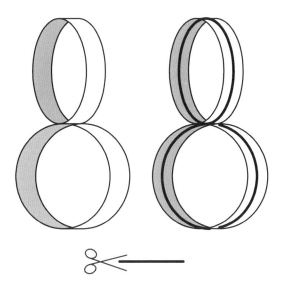

Glue together two perpendicular regular bands (**not** Möbius bands) as shown here. Now cut each through the center line.

What do you get?

5.5. *Perpendicular Band and Möbius Band*

Level of Difficulty: ★
Paper: ▭

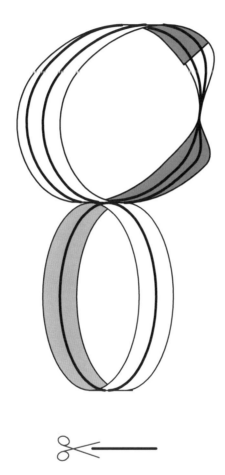

Glue together a Möbius band and a simple band perpendicular to it. Now, cut the Möbius band first, along a line a third of the distance from the edge of the band (not the center). When done, cut the other band.

What do you get?

5.6. *Two Perpendicular Möbius Bands*

Level of Difficulty: ★
Paper: ▭

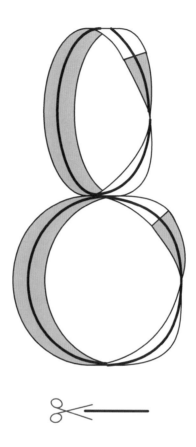

Glue together two Möbius bands, perpendicular to one another. Cut one of the Möbius bands along the center line. When done, cut the other band.

What do you get?

5.7. *Band and S Strip*

Level of Difficulty: ★
Paper: ▭

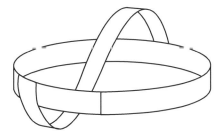 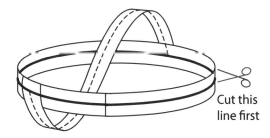

Make a simple band, and glue to it one end of another strip of paper. Take the strip, thread it through the hole in the band from underneath and then glue it to the other side of the band from the top. Cut through the center line of the regular band (thick line) and then cut through the center line of the attached strip (dashed line).

What do you get?

More Strip Puzzles

5.8. *Strip to Pentagon*

Level of Difficulty: ★
Paper: ▭

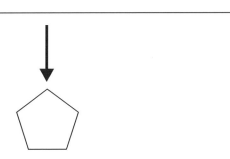

Create the shape of a regular pentagon by folding a long strip of paper.

5.9. *Strip to Hexagon*

Level of Difficulty: ★
Paper: ▭

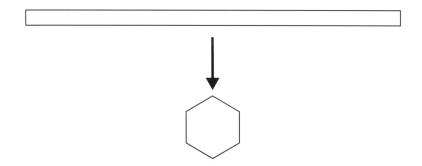

Create the shape of a regular hexagon by folding a long strip of paper.

Yossi Elran suggested the following puzzle based on one of the lesser known characteristics of the Möbius strip.

5.10. *Knotted Strip*

Level of Difficulty: ★★
Paper: ▭

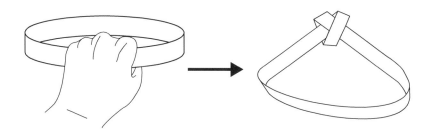

Make a band out of a strip of paper. **Tie a knot in the band without cutting the band open** (that is, without cutting the band along its width)!

> **Hint:** What kind of a band is needed to begin with?

We saw previously how to make a cube from a strip of seven squares. Here is our 'twist' on that puzzle.

5.11. *Strip to Cube*

> Level of Difficulty: ★★
> Paper: ▭

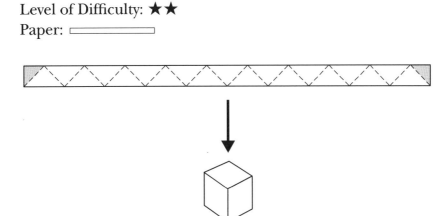

With a strip of paper, fold right-angled triangles out of it (follow the crease pattern in the diagram above). **Now fold a cube, using only existing creases.**

5.12. *Strips to Star of David*

> Level of Difficulty: ★★
> Paper: ▭

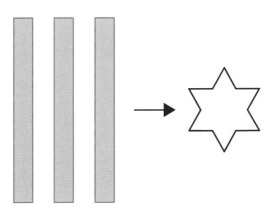

Use three strips of paper to create a Star of David.

This puzzle was introduced by Ivan Moscovich.

Chapter 5: Solutions

5.1. *Möbius Center Cut*

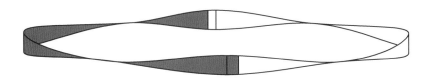

Surprisingly, you do not get two bands, but one, with two sides, and half-twisted four times. By monitoring the process of making the band, it is evident why. Using colors to distinguish between the two halves, you can see that the colors become completely separated, except for the two points of connection.

Follow-Up Activity:

If you cut the new band once again through the center line, what do you get? Try it yourself.

5.2. *Möbius Near-Edge Cut*

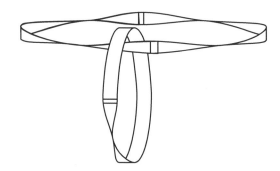

The result is, once again, surprising. The lower cut line rises above the center after the half-twist, continuing at this level for a whole cycle before returning to the lower level after the half-twist and then returning to the starting point. If we take a look at what to expect, it is easy to see that the central part is just a narrower Möbius band. Interwoven into it, there is a double-sided, four half-twisted Möbius band (the same as what we got when we cut through the center of a Möbius band). When we finish the cut, we find that we get two bands locked together — one of them, a half of the length of the other! Can you explain why?

Follow-Up Activity:

Take a strip of paper and half-twist one edge twice before joining the ends. Cut it through the center line. What do you expect to get?

Half-twisting twice ensures that every half (upper or lower) is connected to the same half on the other edge, so that we get two interlocked bands. Because the cut meets at the center point after only one cycle, the bands are equal in length and width.

5.3. *Double Möbius*

The result is, strangely enough, four times a half-twisted band. Since the number of half-twists is even, this is not a one-sided Möbius band! Cut it along its center line to see if we are right.

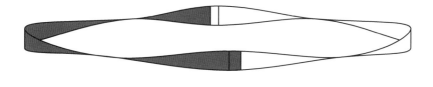

5.4. *Two Perpendicular Bands*

The result is a square frame. It is easy to see why if you cut the bands across the width at the top and bottom of the construction. The result is a cross. The four right angles of the cross reassemble when glued back and cut to create the four corners of the square frame.

5.5. Perpendicular Band and Möbius Band

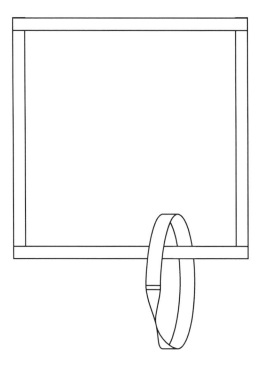

You get a frame interlocked with a Möbius band, one third in width of the original one. It's actually the center part of the original Möbius band.

Unlike before, the order of cutting the bands actually matters. What happens if you first cut the regular band? Why?

If you first cut the regular band, the Möbius band first becomes a strip, and there is no continuity anymore. This takes us back to the puzzle with two regular bands glued together, and the result is the same.

5.6. *Two Perpendicular Möbius Bands*

The result is heartwarming! You get two heart-shaped bands interlocked. Do not be surprised if you get two separated shapes. This works only if your Möbius bands are made in two different orientations. Make sure to rotate one edge clockwise before you glue it; and the other rotated counterclockwise.

Follow-Up Activity:

For further exploration, try and glue together two Möbius bands, **parallel** to one another. Cut one of the Möbius bands along the center line. When done, cut the other band. What do you get?

5.7. *Band and S Strip*

Strangely enough, the result is just the same as the two regular bands glued perpendicularly: a frame. Cutting the black line first creates a strip — a band attached to each end, just like the first cut in the other challenge.

5.8. Strip to Pentagon

Tie a knot in the strip, flatten the flaps and tighten. The knot will form a pentagon.

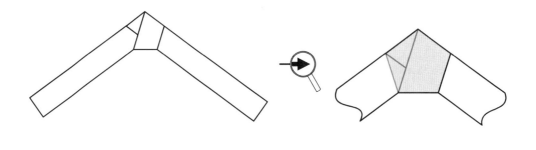

5.9. Strip to Hexagon

Make a double knot, flatten and tighten. A hexagon will be formed.

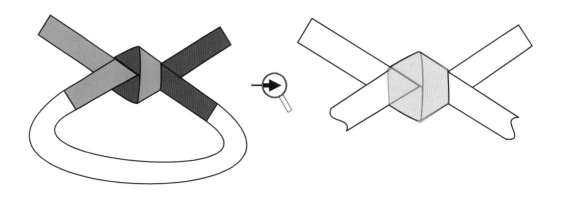

5.10. Knotted Strip

The trick lies in the preparation of the paper band before you start with the cutting. The paper band has to be half-twisted three times. When cutting along the center line and opening up, a band with a knot in it is created.

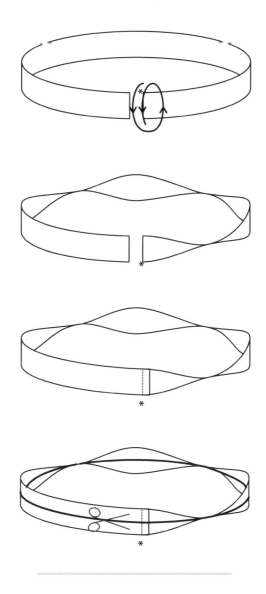

5.11. *Strip to Cube*

Unlike the 1×7 strip (Puzzle 3.1), this strip needs 17 triangles to create a cube. A cube made out of 90° triangles needs only 12 separate triangles (every face needs two, and there are six faces), but making it out of a single strip causes overlappings. **Can you find a way to do it in a shorter strip?**

To get a cube, we need a strip of 17 full isosceles right-angled triangles; and two half-sized ones, on the edges.
Follow the mountain and valley fold lines. Note that most of the folds create a right angle, while two are 180° turns, hence the paper overlaps itself.
The grayed triangles are joined together (each one to the same gray level).

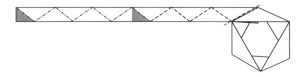

Starting from the left, after six folds of 90°, you get a closed ring. When the gray triangles meet, you need to perform a 180° valley fold.

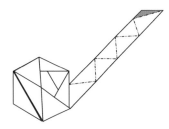

After three more folds, you get to this step. Keep on folding according to the crease pattern, following the valley or mountain directions.

The cube is fully covered with the 17 triangles. The gray ones on the edges are added to help you glue or hold it together.

5.12. *Strips to Star of David*

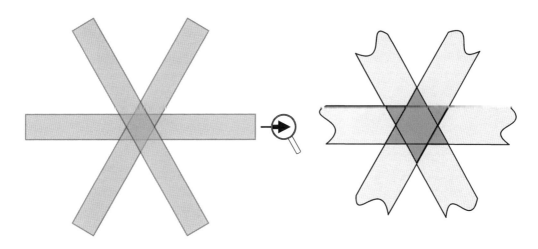

The Star of David is hidden, and can be found only if brightly lighted — from behind!

Overlap the three strips, at a 60° angle and look into the bright sky while holding the three strips together. You can see that the center, where all three strips meet, is darker, and the silhouette forms the hidden star.

Chapter 6

Flexagons

Flexagons are very much related to Möbius strips. Whereas the Möbius strip has less than two faces — it is one-sided — flexagons are 'flat' paper constructions (polygons) that have *more* than two faces. The faces can be 'flexed' so that the different faces come into view. Flexagons are named according to the final shape of the polygon and the number of faces. A tetraflexagon is a flexagon with four sides — a square or rectangle. A tetra-tetraflexagon has four faces, a hexa-tetraflexagon has six and an octa-tetraflexagon has eight. One of the most basic flexagons is the tri-hexaflexagon. It has three faces. This nomenclature is by no means complete and in some cases even contradictory! Each flexagon has its own characteristics, for example, two tetra-tetraflexagons can be different in many aspects, even though they are both four-faced rectangles.

Many people around the world now play with and study flexagons, since they were first invented by Arthur Stone in 1939 and later popularized by Martin Gardner. Some classic references can be found at the end of this book, including the seminal works of Les Pook, David Mitchell, and Hilton and Pedersen. There has been a renewed interest in flexagons in recent years and many more flexagons are being discovered every day. Ann Schwartz from Manhattan probably has one of the most diverse collections of flexagons and leads an internet group on the subject. Together with Yossi Elran, they are investigating the connection between flexagons and Möbius strips and revising flexagon nomenclature.

Flexagons are fascinating, but it can be difficult to actually fold an initial sheet of paper to get them in the first place! Here we pose the construction of flexagons as activities rather than puzzles.

6.1. 2 × 2 Tritetraflexagon

Level of Difficulty: ★

Paper:  Rectangle

Fold the sheet of paper to get a 2 × 2 square flexagon with three faces, so that each face shows one of the digits, three times.

How to make the 2 × 2 flexagon:

Cut out this shape from a 2 × 5 rectangle. Mark the numbers, and turn over.

Complete the numbers and turn over.

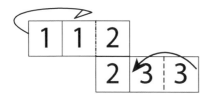

Fold the left flap [1, 1] to the back, and fold the right flap [3] to cover the [3] left to it.

This brings all the [2] squares to the front.

Use tape to hold the right [2]'s together.

How to flex:

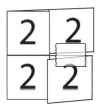 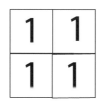 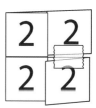

Start with the [2]'s page facing up. Turn over.

The other side has all the [1]'s.
Turn over.

Back to the [2]'s side.

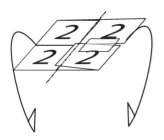  

Fold both sides backward.

In process.

In process.

  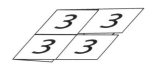

Till it is flat.
Now open it like a book pointing up. On the right take only one layer, while on the left take two layers.

In process.

The [3] page is revealed. You can repeat this process to get all possible pages, rotating from [1]'s to [2]'s and [3]'s.

6.2. 2 × 2 Hexatetraflexagon

Level of Difficulty: ★

Paper: 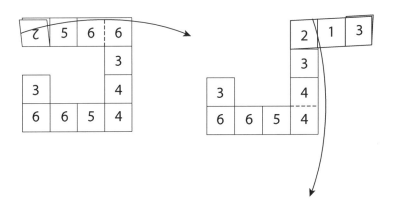 Square

Fold the sheet of the cut-out paper to get a 2 × 2 square flexagon with six faces, so that each face shows one of the digits, six times.

How to make:

*4	5	6	6
4			3
3			4
6	6	5	4

5	2	1	3*
1			2
2			1
3	1	2	5

4	5	6	6
4			3
3			4
6	6	5	4

Make a 4-by-4 grid, and cut away the center 2 × 2 square. Add a cut between the [2] and the [1] on the right side.

Valley fold [4] on [4].

Cover the [6] with the [6], by valley folding to the right.

Cover the [4] with the [4], by valley folding the right top section downward.

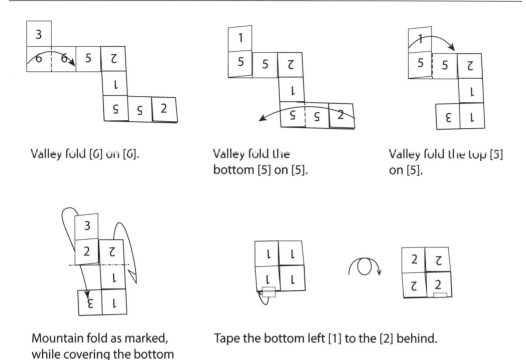

Valley fold [6] on [6].

Valley fold the bottom [5] on [5].

Valley fold the top [5] on [5].

Mountain fold as marked, while covering the bottom [3] with the upper [3].

Tape the bottom left [1] to the [2] behind.

How to flex:

Apply the same technique as you did with the *tritetraflexagon*. Try and find all six faces.

6.3. 2 × 2 Hexatetraflexagon II

Level of Difficulty: ★

Paper: Square

Fold the sheet of paper to get a 2 × 2 square flexagon with six faces, so that each face shows one of the digits, six times. Unlike the previous flexagon, which was made from a strip, this is made from a closed band, and requires a more complicated manipulation.

How to make:

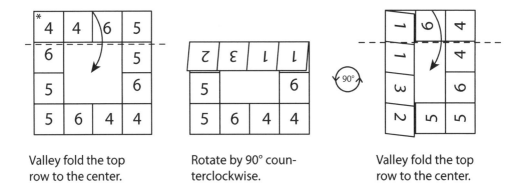

Make a 4-by-4 grid, and cut away the center 2 × 2 square.

Valley fold the top row to the center.

Rotate by 90° counterclockwise.

Valley fold the top row to the center.

Chapter 6: Flexagons

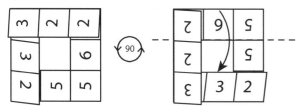

Rotate by 90° counterclockwise.

Valley fold the top row to the center.

Rotate by 90° counterclockwise.

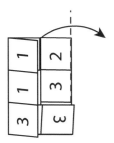

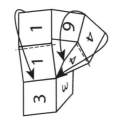

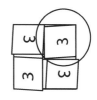

Open the flap [2, 3] to the right. The model is not flat!

Exposing the [6] and [4], tuck the upper opened row in, so [4] covers [4] and [1] covers [1] below it.

This brings the hidden [3] forward! Your flexagon is ready!

How to flex:

Apply the same technique as you play with the *2 × 2 flexagon*. Try and find all six faces.

6.4. *2 × 3 Tetratetraflexagon*

Level of Difficulty: ★

Paper:

Fold the pre-cut sheet of paper to get a 2 × 3 rectangular flexagon with four faces, so that each face shows one of the digits, four times.

How to make:

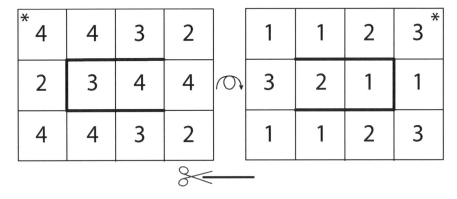

Make a 4-by-4 grid, and cut away one row. Cut the marked 1 × 2 rectangle in the center, leaving one side intact.

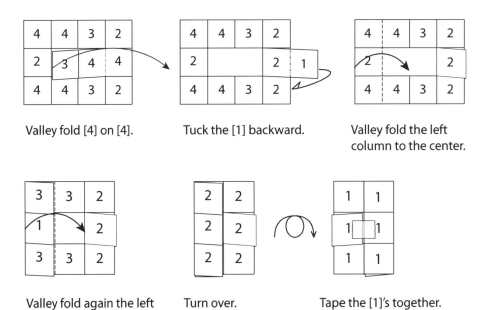

Valley fold [4] on [4].

Tuck the [1] backward.

Valley fold the left column to the center.

Valley fold again the left column to the next fold line.

Turn over.

Tape the [1]'s together. The flexagon is ready to use!

How to flex:

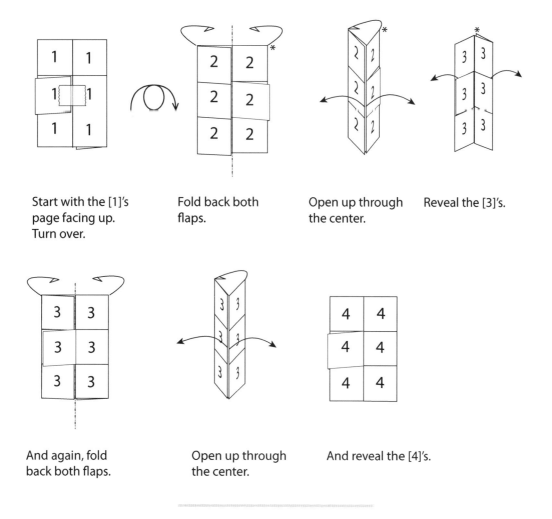

Start with the [1]'s page facing up. Turn over.

Fold back both flaps.

Open up through the center.

Reveal the [3]'s.

And again, fold back both flaps.

Open up through the center.

And reveal the [4]'s.

Hexaflexagons

Hexaflexagons are multi-faced flexagons in the shape of a hexagon (as opposed to the rectangular or square flexagons we saw before). First, try and construct a trihexaflexagon with three faces, each colored a different color.

6.5. Trihexaflexagon

Level of Difficulty: ★
Paper: ▭

Fold the strip of paper to get a hexagonal flexagon with three faces, each face made of six equilateral triangles. All the triangles on each face have to be colored with the same color, and all the faces have to be colored differently.

Hexaflexagons:

Use a long strip and divide it into 10 equilateral triangles. Copy this pattern.

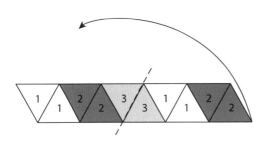

Valley fold along the marked crease line.

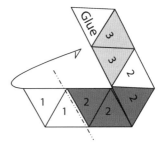

Mountain fold along the marked crease line.

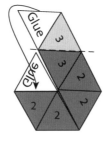

Valley fold along the marked crease line. The two tabs marked with **Glue** are matched perfectly in the text location. Apply glue.

Your flexagon is ready!

Chapter 6: Flexagons

How to flex:

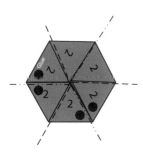

Start with the [2] face. Note the exact location of the glued flaps and the valley and mountain folds, accordingly!
Hold at the black dots.

Pinch two mountains, and make sure the third is mountain-folded, too.

Continue to pinch, until all flaps are in pairs. You can see in the middle a space is created.

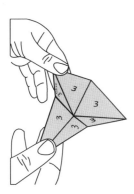

Keep on opening this space, revealing a new face, made of [3]'s.

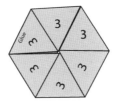

You have discovered a new face! Keep on doing the same to uncover all three faces!

Here are some more complicated flexagon constructions.

6.6. *Hexahexaflexagon*

Level of Difficulty: ★
Paper: ▭

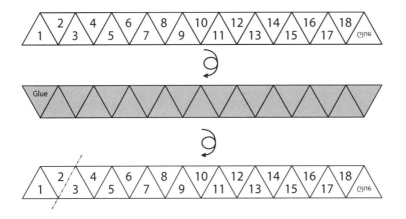

Mountain fold the long right flap; between [2] and [3].

Valley fold eight more times, to get a 10-triangle strip. Turn over.

Mountain fold the short end between [4] and [7].

Mountain fold the top end between [11] and [12] and bring the last flap **above** [1].

Chapter 6: Flexagons

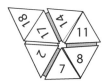 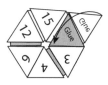 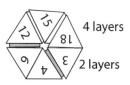

To get this.
Turn over.

Fold and glue the flaps.

Note the number of layers on every side.

How to use:
Operate the hexahexaflexagon just like the simple hexaflexagon. However, you will get many more positions.

6.7. *Flexagon Rotor*

Level of Difficulty: ★

Paper: 1 : 4

This model can be rotated forever!

Hold the upper flaps and turn them outward, to reveal two more flaps underneath. You can rotate them as well, just to uncover more flaps waiting, and so on, endlessly.

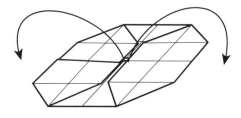

How to make the flexagon rotor:

Start with a 16 × 4 grid. Cut away three columns to get a 13 × 4 grid.

Fold both edges to the center.

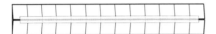

Tape the flaps to each other. Turn over.

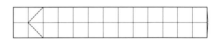

Crease the marked fold lines. Fold and unfold along the crease line to make later folding easier.

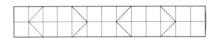

Repeat all over.

Fold the left flap to a standing position (90° to the surface). Next image is from an isometric point of view.

Squash the corners, till they meet in the center, by using the existing crease lines only.

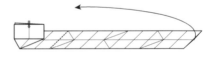

Fold the right flap to a standing position (90° to the surface).

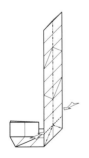

Squash again.

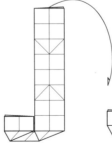

Flatten the right flap on the table.

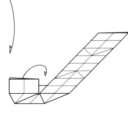

And the left one, too.

Turn over.

Chapter 6: Flexagons

Raise the long flap.

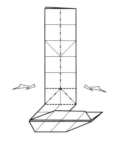

Squash again.

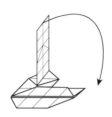

Flatten again.

Turn over.

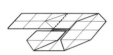

Raise the flap.

Squash again.

Insert the end flap into the first one we folded.

Add a piece of tape to hold it together.

The model is ready!

To activate, hold the two inner flaps and turn them outward.

Two new flaps appear, and you can flip them out, too …

… to reveal two more flaps; and again, and again.

6.8. *Tetrahedron Hexaflexagon Rotor*

Level of Difficulty: ★

Paper: 1 : 2

Tomoko Fuse suggested this 3D version of a flexagon. It shows three different faces, and can rotate inward endlessly. Her version is made of three sheets. Ours is a single sheet version.

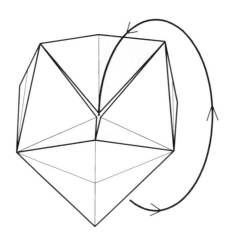

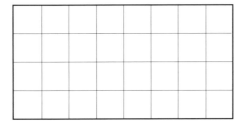

Divide a 1 : 2 rectangle into 4-by-8 squares.

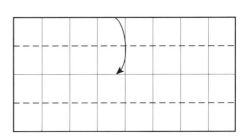

Fold both top and bottom edges to the center line.

Like this. Turn over.

Valley fold a diagonal of two squares. Unfold.

Chapter 6: Flexagons

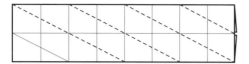

Repeat as marked.

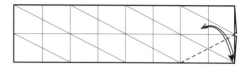

Now continue with the other diagonal of the last rectangle.

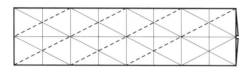

Complete all the rest.

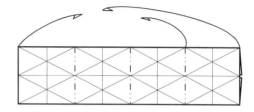

Use the three existing vertical fold lines to fold the paper into a cube-like shape.

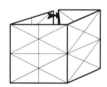

Insert the (original) left edge into the right edge. (If one edge is a little bigger than the other, insert the small edge into the bigger one.)

Insert it all the way to the end. Next image is enlarged.

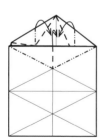

Tuck in the three midsections of the upper edges.

Now tuck in the three marked corners, until the three meet in the center. Be firm!

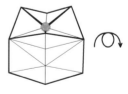

To get this. Turn the model upside-down.

  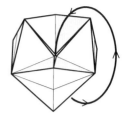

Tuck in again the three midsections.

And push inward the three marked corners.

The finished model is actually made of six connected tetrahedra. You can rotate the model inward endlessly.

 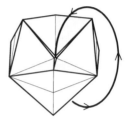

While rotating, you will see the open faces. You can tape them for better performance.

The rotating tetrahedron hexaflexagon is ready!

Chapter 7
Fold and Cut

Fold-and-cut puzzles are construction puzzles which involve both cutting and folding. The extra 'twist' that the cut adds to the puzzle makes this kind of puzzles more complicated, even when dealing with very few cuts and folds.

7.1. *Impossible Object*

Level of Difficulty: ★

Paper:

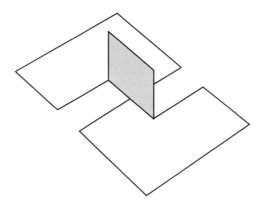

Cut a rectangle sheet of paper and fold to create this seemingly impossible object.

If the two edges (left and right) of the paper are joined together, the result is a Möbius band!

7.2. *How Many Pieces?*

Level of Difficulty: ★

Paper:

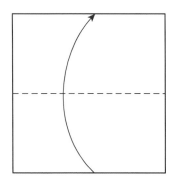
Fold in half.

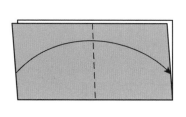
And again.

Cut along the line. How many pieces will you get?

Fold a piece of paper bottom to top, and then fold the result from left to right. Cut a straight line from top to bottom as shown above. **How many separate pieces are there?**

Add a third fold and cut. **How many separate pieces are there now?**
Can you solve this without trying it out on paper first?

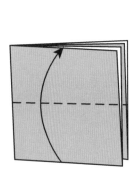
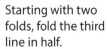
Starting with two folds, fold the third line in half.

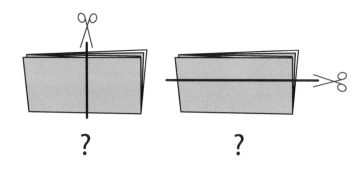
Cut either in perpendicular to the last fold line, or in parallel to it. How many pieces will you get in each case?

Single Fold; Single-Cut Puzzles

David Goodman suggested the following puzzle based on his favorite theme, the Star of David, and two other of his inventions.

7.3. *Star of David*

Level of Difficulty: ★★

Paper:

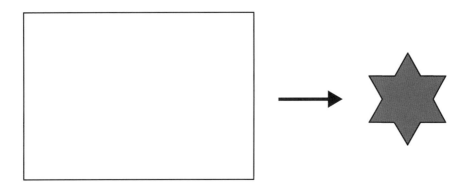

Fold a single sheet of paper, and make a single cut. Rearrange all the pieces so that you can see a Star of David.

7.4. *Maximum Length*

Level of Difficulty: ★

Paper:

By using one fold and a single straight line cut, what is the maximum length you can cover by putting the three pieces in a straight line?

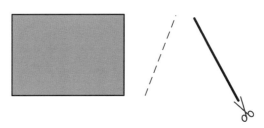

MAX length

7.5. Silhouettes

Level of Difficulty: ★ to ★★

Paper:

If you fold a piece of paper once and then make a single cut along a straight line, you can get either two or three pieces. (Can you explain why?)

This is a tangram-like puzzle, where you first have to find the right pieces and then arrange them correctly.

Here are four different puzzles. Each needs a different location for the fold line and the cut line.

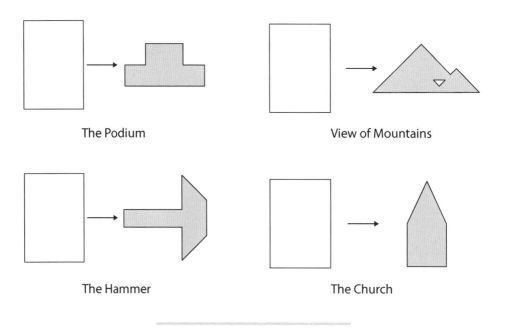

The Podium

View of Mountains

The Hammer

The Church

Multiple Folds; Single-Cut Puzzles

Martin and Erik Demaine (a father-and-son team of mathematicians working at MIT) developed a proof and an algorithm with which any polygon can be made by folding a sheet of paper and then making a single cut with scissors.

The idea is to fold the paper in such a way that all the edges of the polygon align. Cutting this line with scissors and unfolding it reveals the desired polygon.

This concept creates a fertile ground for many puzzles. We tried to highlight the best ones, but you can literally challenge yourself with any shape you want. Demaine and Demaine provided paper templates for a swan and many other shapes. In theory, of course (see reference), any polygon can be made. However, many polygons require so many folds, that, in practice, they are impossible to make with a regular sheet of paper. Furthermore, there is still no proof that Demaine and Demaine's algorithm is the optimal (i.e. with the least possible number of folds).

We present here some of our own 'fold-and-cut' puzzles, and some classics by Gerald Loe and Will Blyth.

7.6. *Fold-and-Cut Square*

Level of Difficulty: ★

Paper:

* Size of square is not to scale!

Fold as much as you like, so a single cut will create a square with a squared hole.

7.7. Fold-and-Cut Star of David

Level of Difficulty: ★★★

Paper:

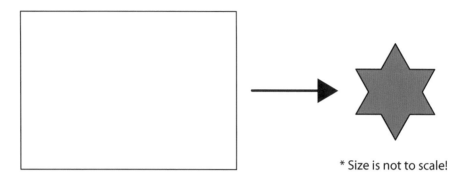

* Size is not to scale!

Cut a Star of David from a rectangle by folding the paper and then cutting once along a straight line.

7.8. Fold-and-Cut Hollow Star of David

Level of Difficulty: ★★★

Paper:

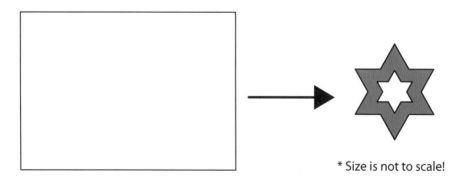

* Size is not to scale!

Cut a hollow Star of David from a rectangle by folding it and then cutting once along a straight line.

7.9. Fold-and-Cut Cross

Level of Difficulty: ★★

Paper:

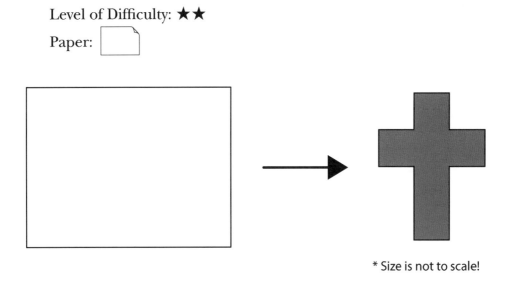

* Size is not to scale!

Cut a cross from a single sheet, by folding it and then cutting once along a straight line.

7.10. Fold-and-Cut A to Z

Level of Difficulty: ★★ to ★★★★

Paper:

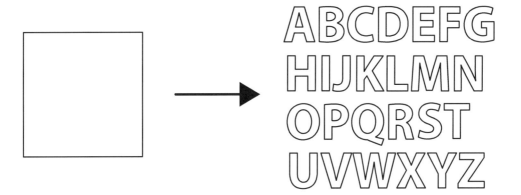

Make the complete ABC; all by folding and making a single scissor cut.

7.11. *Strip to a Square*

Level of Difficulty: ★★ to ★★★

Paper: 1 : 5

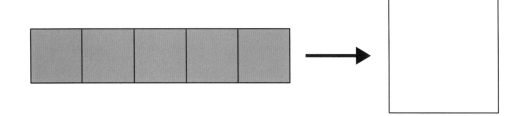

Sarcone and Waeber suggested this puzzle.

Fold a rectangular strip with aspect ratio 1 : 5 and then make two cuts, each along a straight line, so that you can form a square from the resultant pieces.

Hint: To solve this puzzle, you may need to fold the paper before making the cut.

7.12. *Impossible Flap*

Level of Difficulty: ★★

Paper: 3 : 4

This is a classic invented by Robert Neale and popularized by Karl Fulves.

Cut a 'window' flap in the center of a sheet of paper (to make it easy, use a 3×4 grid). **Hold the flap from beneath the sheet as shown here, and then maneuver it to a new position above the plane of the sheet without letting go of the flap.**

Chapter 7: Fold and Cut

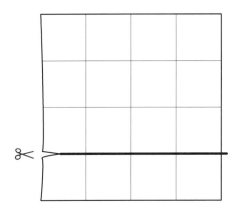

For this puzzle you need a 3 × 4 rectangle. Make a 4 × 4 square and cut away the bottom row. Make sure all lines are creased.

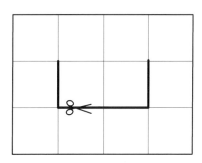

Cut along the thick lines, making an inward flap the size of two squares.

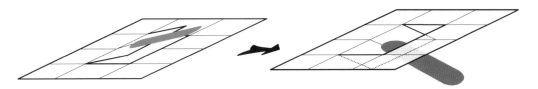

Hold the inner flap with your fingers, so the rest of the paper is underneath them. By folding only along existing crease lines, bring the rest of the paper to be above your fingers!

Chapter 7: Solutions

7.1. *Impossible Object*

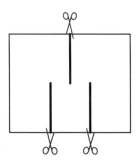

Cut three slits in the paper as shown.

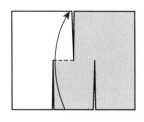

Fold bottom edge to the top, but move only the dark area!

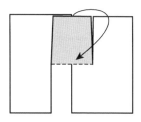

Bring the free flap (marked in gray) to stand perpendicular to the table.

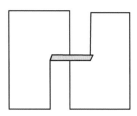

To this position.

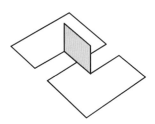

Side view.

7.2. *How Many Pieces?*

You get three pieces if you fold the paper twice.

If you add a third fold, the answer is "that depends…". Unlike the first step, where it doesn't matter which way you cut it, the direction of the third cut does make a difference.

If you cut perpendicular to the third fold, you effectively undo the third fold and the answer remains as three pieces.

If you cut parallel to the third fold, you will get five pieces.

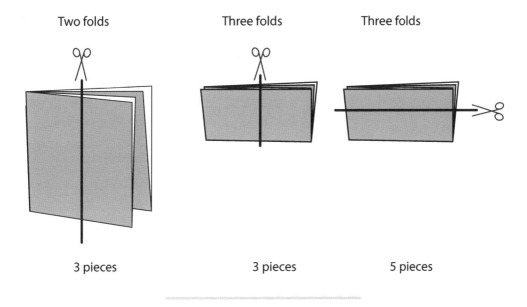

7.3. Star of David

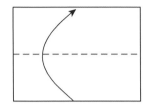

Fold in half.

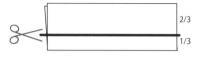

Cut along the one third line.

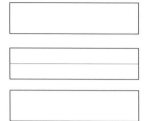

Spread to get three strips.

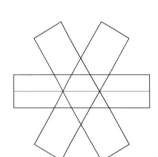

And this is something we already solved ...

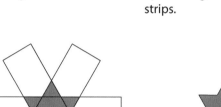

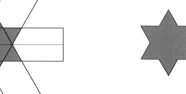

And here is another solution:

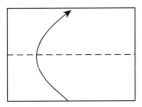
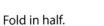
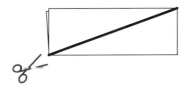
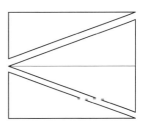

Fold in half.　　　　　　　Cut along the diagonal.　　　　Spread to get three parts.

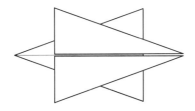

And rearrange to form the Star of David.

7.4. Maximum Length

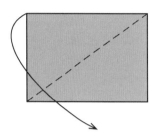
Fold along the diagonal.

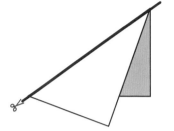
Cut as close as you can to the fold line.

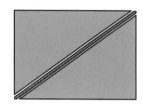

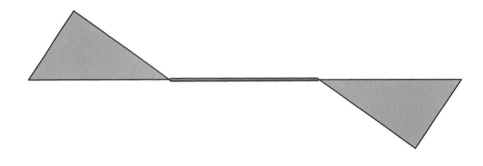

7.5. Silhouettes

The Podium

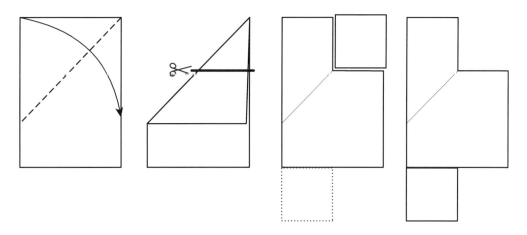

View of Mountains

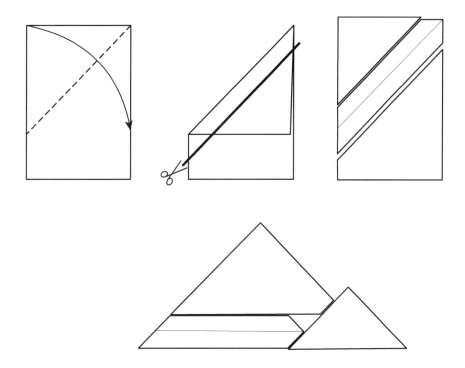

The Hammer

The Church

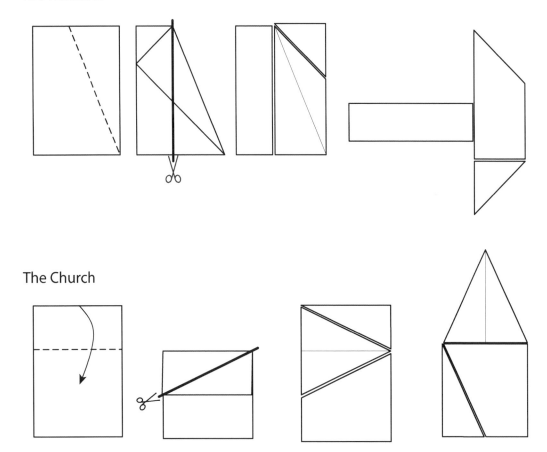

7.6. Fold-and-Cut Square

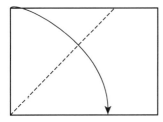

Fold the left edge to the bottom edge.

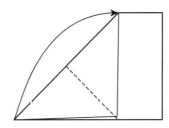

Fold the left new corner to the new obtuse angle on the top.

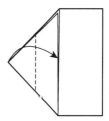

Fold the left corner to the right, to meet the edge of the top layer.

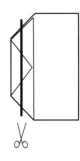

Cut along the marked line.

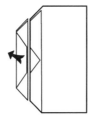

Open up the left part.

A square in a square appears!

7.7. Fold-and-Cut Star of David

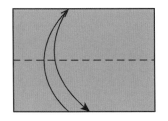

Fold bottom to top and unfold.

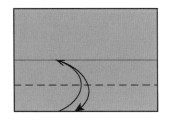

Fold the lower edge to the center line and unfold (to mark the quarter line).

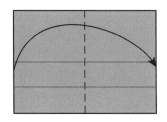

Fold in half from left to right.

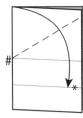

Fold the upper left corner to the quarter line (∗), and make sure the fold line starts at the (#) mark.

Fold the bottom left corner to the upper edge. A 60° angle is made!

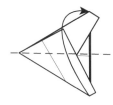

Fold in half with the bottom covering the top.

Cut along the thick line.

Unfold the left triangle.

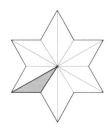

A Star of David is formed!

7.8. Fold-and-Cut Hollow Star of David

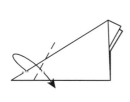
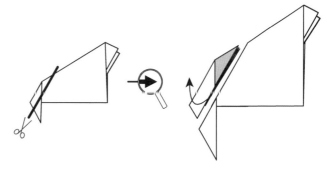

Repeat the folding process of the previous solution. Add a valley fold as marked (at 60°).

Cut along the thick line.

Unfold the left part (shaded in gray). Throw the rest away.

Unfold all.

The hollow Star of David.

7.9. Fold-and-Cut Cross

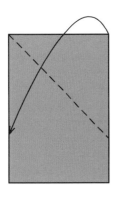

Fold the top right corner to the left side.

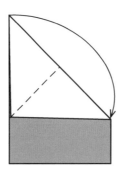

Fold the top left corner to the right side.

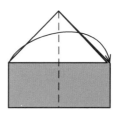

Fold in half; left to right.

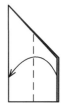

Fold in half again; right to left.

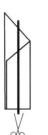 

Cut along the thick line, and unfold the gray part.

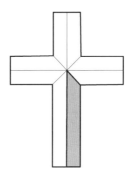

Here is the cross.

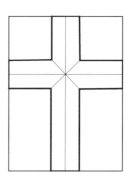

This is the equivalent puzzle — bring all the thick lines together.

7.10. Fold-and-Cut **A** to **Z**

Here are some examples for the solutions. We leave you the rest to solve on your own!

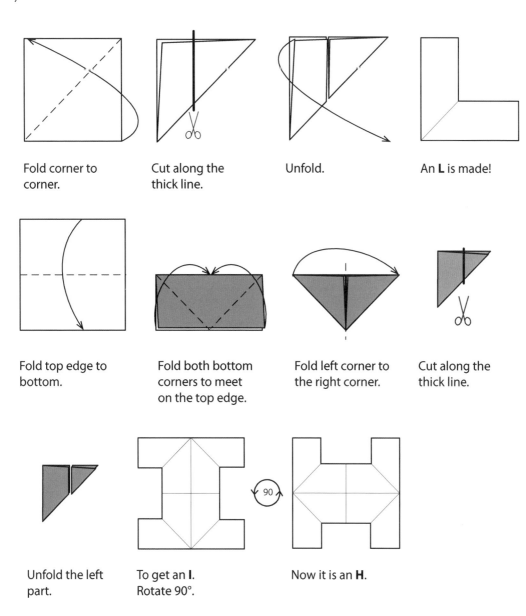

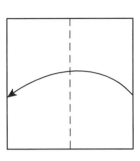
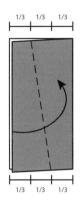
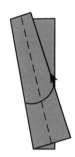

Fold right edge to left.

Mark the thirds, and fold as shown.

Fold left edge to the right edge of the front flap.

Cut along the thick line.

Unfold the main part.

An **M**, a **W** or an **E** emerges.

7.11. *Strip to a Square*

A five-unit solution
This is the original solution by Sarcone and Waeber.

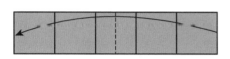 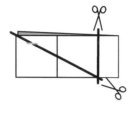 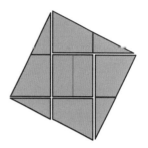

Fold edge to edge.

Cut along the thick lines.

Rearrange the five pieces into a square.

A four-unit solution

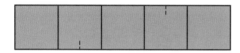 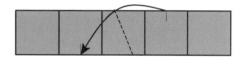

Pinch the center of the second and fourth squares. The second at the bottom edge, and the fourth at the top edge.

Fold pinch mark to pinch mark.

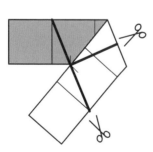 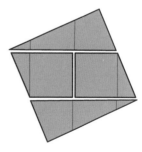

Cut along the thick lines.

Rearrange the four pieces into a square.

A four-unit solution
by Dominique Ceugniet

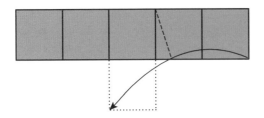
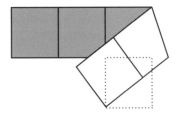

Imagine an extra square under the middle one. Bring the bottom right corner to meet the bottom left corner of this square.

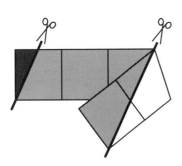

Cut along the thick lines.

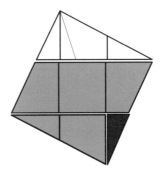

Rearrange the four pieces into a square.

7.12. Impossible Flap

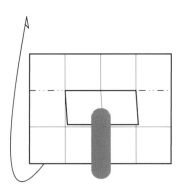

Hold the inner flap with your fingers above the rest of the paper. Fold the big flap backward.

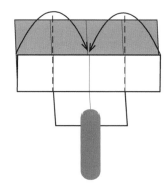

Fold the sides to the center.

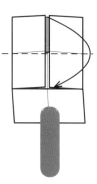

Fold the top part down.

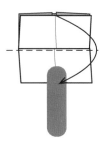

Fold the upper flap only (the white one) down.

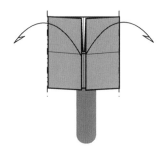

Open the two flaps to the sides.

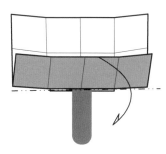

Fold the gray flap down.

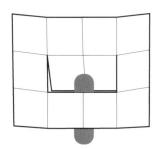

Your fingers hold the flap now, underneath the rest of the paper!

Chapter 8
Just Cutting

Paper puzzles are fascinating. Up to now, most of the puzzles involved some kind of folding. However, there are some puzzles, where all you need to do is cut!

Here are some examples, beginning with a few of Robert Neale and Henry Dudeney's classics.

8.1. *A Coin through a Hole*

    Level of Difficulty: ★

    Paper:

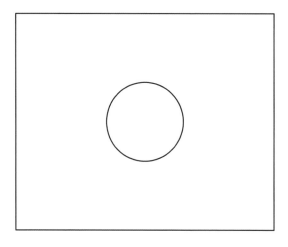

Make a coin go through a circular hole with a diameter 2/3 of the coin's diameter, without tearing the paper.

8.2. *Hole through a Card*

Level of Difficulty: ★ to ★★

Paper:

Cut a hole in a postcard large enough for you to walk through it!

This is a classic. It is highly surprising that you can cut such a large hole from such a small piece of paper.

8.3. *Hole through a Card with Two Cuts*

Level of Difficulty: ★ to ★★

Paper:

Here is an original variation on the hole through a card trick. In the original puzzle, there is no limit to the number of cuts you make. Here every cut counts!

Cut a hole in a postcard large enough for you to walk through it, using only two continuous cuts.

8.4. *House to Square*

Level of Difficulty: ★★★

Paper:

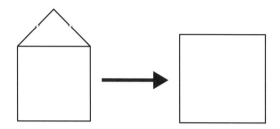

Cut the shape along two straight lines so that the three resultant parts reassemble into a square.

Note: The roof is a quarter of the house.

There are no foldings (except for finding reference points) or overlapping needed to solve this puzzle.

8.5. *Cut a Fan*

Level of Difficulty: ★★

Paper:

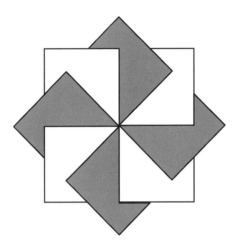

Recreate the shape with a minimal number of sheets of paper. No folding is allowed, but cutting is!

8.6. *Interlocking Rings*

Level of Difficulty: ★★★

Paper:

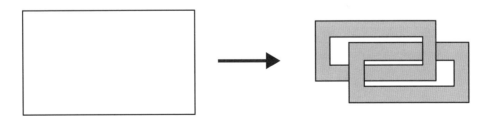

Cut a chain of two interlocking closed rings from a single sheet of paper.

8.7. *Cube Net to a Square*

Level of Difficulty: ★★★

Paper:

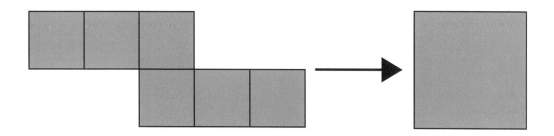

Cut a 'cube net' shape along two straight lines so that the resultant parts reassemble into a square.

You can move the parts after the first cut.

From all 11 possible nets, the only solution we know is the one above.

We encourage you to look for more, and if you do find some, please let us know!

Chapter 8: Solutions

8.1. *A Coin through a Hole*

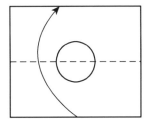

Fold the bottom edge to the top.

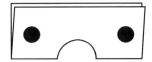

Pinch the paper at the black dots and rotate your wrists inward.

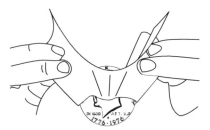

The coin will go through!

8.2. *Hole through a Card*

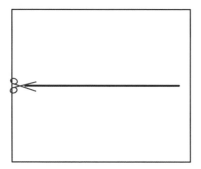

Cut the thick line. Make sure you DO NOT cut to the edges!

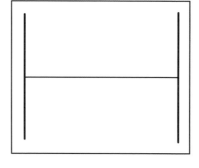

Cut the thick lines. Make sure you DO NOT cut to the edges!

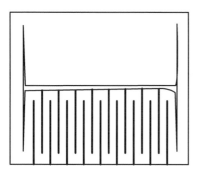

This cutting process makes the bottom side much longer.

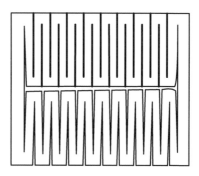

Repeat on the upper side.

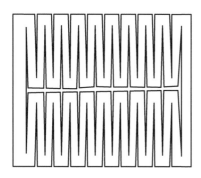

Now you can see the potential …

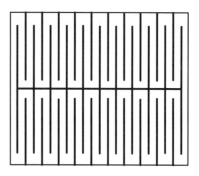

… cutting diagram.

8.3. *Hole through a Card with Two Cuts*

The solution is based on a spiral cut. The first cut creates a very long strip of paper, spiralled inward. The second cut creates a slit in the strip which turns it into a very long band.

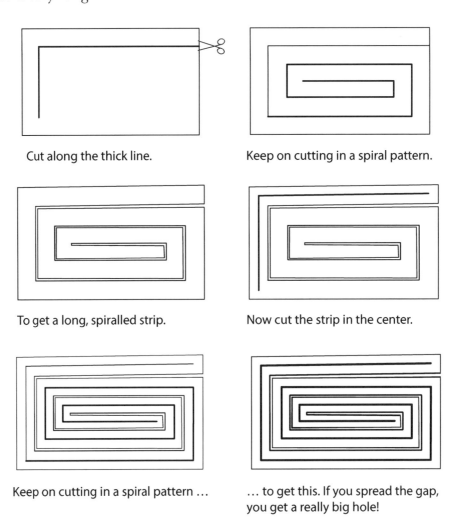

Cut along the thick line.

Keep on cutting in a spiral pattern.

To get a long, spiralled strip.

Now cut the strip in the center.

Keep on cutting in a spiral pattern …

… to get this. If you spread the gap, you get a really big hole!

The full cutting pattern.

8.4. *House to Square*

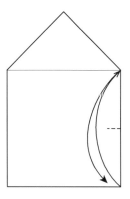

Mark half of the right edge with a pinch.

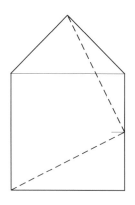

Fold the dashed lines made from the pinch mark to the two corners.

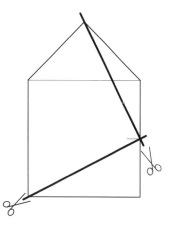

Cut along the creases.

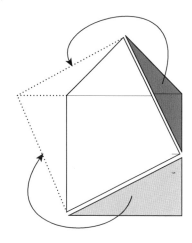

Move the parts as marked.

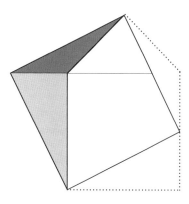

A square is formed.

8.5. *Cut a Fan*

Surprisingly, only two sheets are needed. The smart cut allows to interlock them in a way that looks like you need eight squares, four of each color.

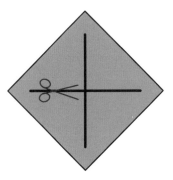

Cut along the thick lines.

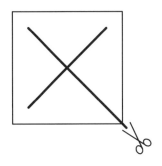

Cut along the thick lines.

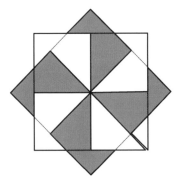

Interlace the two parts
into this shape.

8.6. *Interlocking Rings*

We treat paper as if it was two-dimensional. If this was really true, this puzzle could not be solved. Nothing is only two-dimensional in our world. The way to make the overlapping rings is to use the third dimension by dividing the paper into two layers at the cross sections.

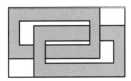

Draw this diagram on a sheet of heavy, thick paper; at least 250 GSM.

Cut out the dark parts.

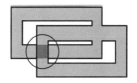

The next step is done on the cross section marked in the circle.

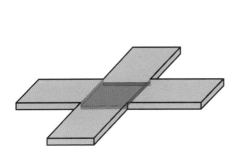

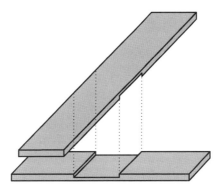

Using the thickness of the paper, gently spread the paper into two layers. You can add pre-cuts, halfway through the paper thickness, before you peel off the top layer. Repeat on the other cross section.

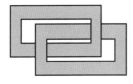

This is the result.

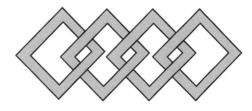

You can make as many units as you want with the same process.

8.7. Cube Net to a Square

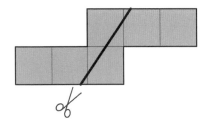

Cut along the thick line.

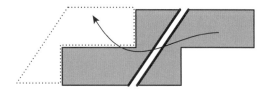

Move the right part as marked.

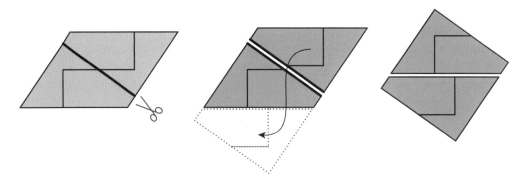

Cut along the thick line.

Move the top right part as marked.

A square is formed.

Chapter 9
Overlapping Paper Puzzles

This chapter is dedicated to Ilan Garibi's overlapping square puzzles. In the following puzzles, all you need are sheets of paper. No folding or cutting is allowed or needed!

9.1. *Overlapping Sheets in a Square — Different Sizes*

    Level of Difficulty: ★★

    Paper: ▢ , Any Size

Arrange square sheets of paper one on top of the other to form a square. **What is the smallest number of sheets needed to ensure that no sheet is fully visible?** There are no other limits to this puzzle.

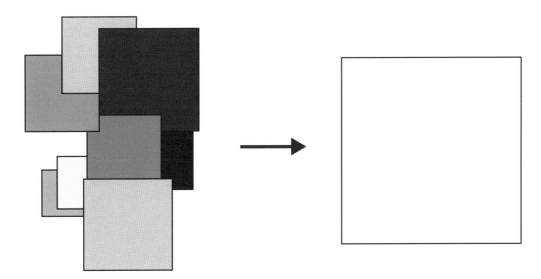

Explanation and counterexample:

This figure shows a counterexample using four different colored square sheets of paper. You can see that there is one sheet which is totally visible, which does not fulfil the requirement of the solution.

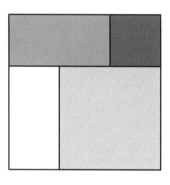

9.2. *Overlapping Sheets in a Square — Equal Sizes*

Level of Difficulty: ★★

Paper:

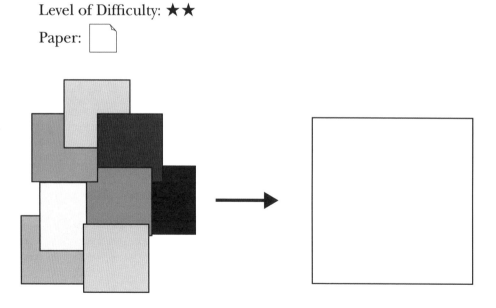

Arrange square sheets of paper, *all of the same size*, one on top of the others to form a square. What is the smallest number of sheets needed if it is required that no sheet is fully visible?

9.3. *Kissing Sheets — All Kiss Together*

Level of Difficulty: ★★

Paper:

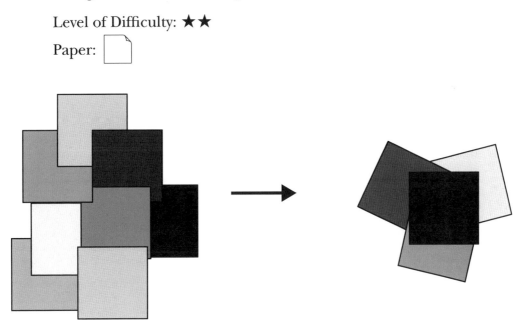

What is the largest number of square sheets of paper, all of the same size, you can arrange together, so that each touches ('kisses') all others? 'Touching' means face to face, only.

9.4. *Kissing Sheets — Couple's Kiss*

Level of Difficulty: ★★★
Paper: ▭

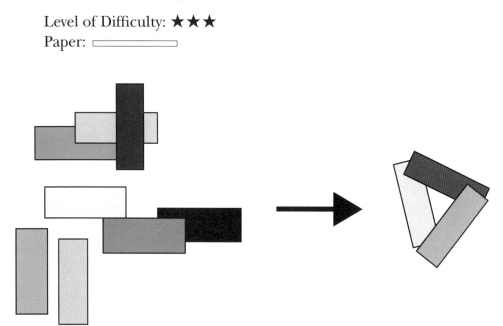

What is the largest number of sheets of paper (without limits on size or shape) you can arrange together, so that each touches ('kisses') all others, and having no more than two overlapping sheets at each point of contact?

Chapter 9: Solutions

9.1. *Overlapping Sheets in a Square — Different Sizes*

Four sheets is the minimal amount. Three that cover each other and the last sheet, large enough to encompass this assemble and placed behind them, solve the puzzle.

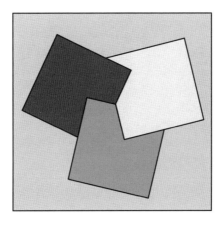

9.2. *Overlapping Sheets in a Square — Equal Sizes*

The minimum number of identical sheets you need is eight. The building block is the mutually overlapping 'plus sign' shape, shown below, made out of four sheets of paper. Add four more sheets for the corners.

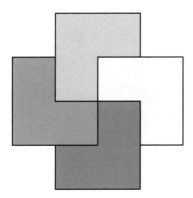

Four sheets to cover each other.

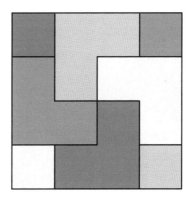

Four more underneath assemble the background square.

9.3. Kissing Sheets — All Kiss Together

The true number of papers required so that each touches all the others is infinite, assuming the sheets have no thickness.

In reality, there will be a limit, of course, due to the fact that the upper sheet at a certain point will be too high to touch the lower one.

9.4. Kissing Sheets — Couple's Kiss

Not allowing more than two sheets to overlap seems to be a major restriction. However, since there are no restrictions on the shape of the paper, the answer is again infinity.

If only square sheets are used you get only three, but by using very long strips of paper, the number increases up to infinity.

Draw a rounded line in the shape of a half-circle. Place additional straight strips tangent to the curve at different points. These points are the 'overlapping' points.

As long as the strips are long enough, they should meet at some point. The reason is simple — at any given point outside a circle, only two tangent lines exist. No third line can meet at the same spot.

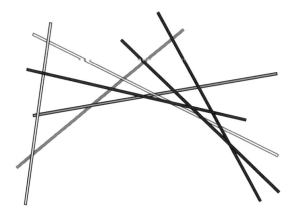

By using very long and narrow strips of paper, you can manage to get an (almost) endless number of sheets of paper, each touching all others, with no more than two sheets meeting at any point.

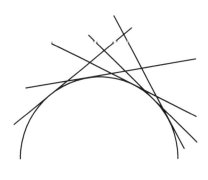

From any point outside the curve, only two tangent lines can emerge.

Chapter 10

More Fun with Paper

In this last chapter, we chose to add a few classical fun things you can do with a piece of paper. We felt that we really couldn't publish a book on paper puzzles without this. To be truthful, there really aren't any 'puzzles' to these challenges, save the fact that you need to actually make them and see them work. Most of these are classics going way back to the 18th century and popularized by Martin Gardner, Karl Fulves and others.

10.1. *Torpedo (Spinning Fish)*

Level of Difficulty: ★
Paper: ▭

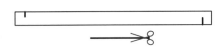

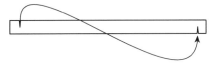

Use a strip of 2 by 20 cm. Make a cut halfway through the width on both sides, 2 cm from the edges. Make sure one is on the top part and the other is on the bottom.

Insert the top cut into the bottom cut, connecting them.

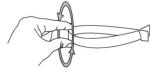

Pinch the center part gently, to make a slight crease.

The spinning fish is ready!

Hold it between your thumb and finger. Raise your hand and give it a spin, while letting it go!

Follow-Up Activity:

You can try this model with different paper sizes and types. Use a wider strip, for example, and see if it spins differently.

10.2. *Rotator*

Level of Difficulty: ★★
Paper: ▭

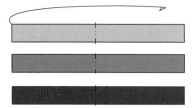

Start with three 1 × 10 strips. Fold all in half.

Like so.

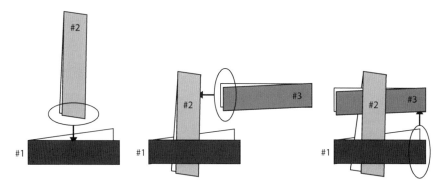

Lay folded strip #1 on the table.
Insert both edges of strip #2 in between the two layers of the first folded strip.

Insert strip #3 in between the two layers of the second strip.

Insert in the same way the edges of strip #1 in between the two layers of strip #3.

Pull each strip by its open side, until all are tightly assembled, and a cube corner is formed. All the angles between every two strips are 90°.

Hold it as shown; raise it high and let go. The rotator will spin while falling down, and keeps on spinning, as it falls on a smooth surface.

10.3. *Helicopters*

Level of Difficulty: ★

Paper: 1 : 3

This is a model from our childhood. It gave us much enjoyment, rotating all the way down from the balcony to the street. Proportions are highly important, and if the wings are too long, the aircraft will go into an uncontrolled dive. You can try different proportions or change the length of the balance, to increase or decrease the weight, and accordingly the stability of the helicopter.

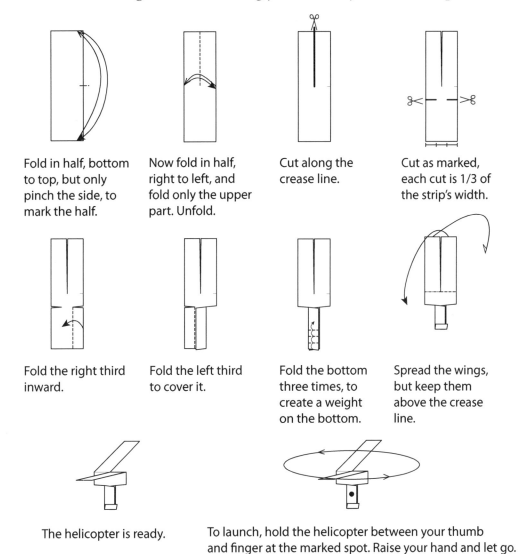

Fold in half, bottom to top, but only pinch the side, to mark the half.

Now fold in half, right to left, and fold only the upper part. Unfold.

Cut along the crease line.

Cut as marked, each cut is 1/3 of the strip's width.

Fold the right third inward.

Fold the left third to cover it.

Fold the bottom three times, to create a weight on the bottom.

Spread the wings, but keep them above the crease line.

The helicopter is ready.

To launch, hold the helicopter between your thumb and finger at the marked spot. Raise your hand and let go. The helicopter will descend and the rotor spins.

Another simpler version for the helicopter uses a much heavier balance.

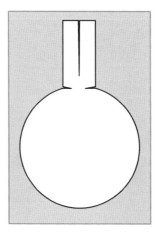

Cut out of a printer paper.

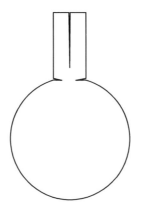

Crumple all the circle, to create a heavy balance.

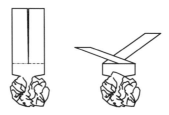

Spread the wings as with the previous model, and your helicopter is ready to fly!

Noise Makers

10.4. *Origami Screecher*

Level of Difficulty: ★

Paper:

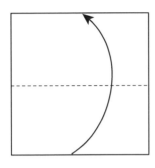　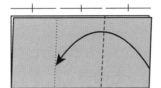　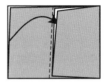

Use a square sheet of paper, 10 × 10 cm. Fold edge to edge.

Divide the paper into three and fold the right third of the paper to the 2/3 line.

Insert the left third of the paper into the pocket created in the previous step.

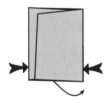　

Shape the lower opening so you can blow inside easily.

Hold between thumb and finger at the black spots, and blow into the opening.

10.5. *Whistle*

Level of Difficulty: ★

Paper:

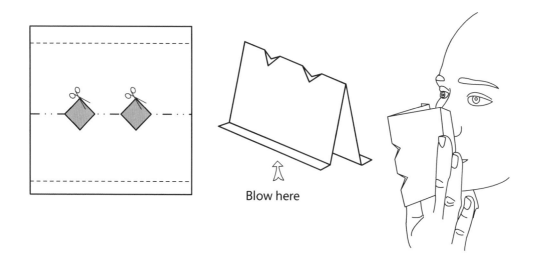

Fold in half to mark the center line. Cut two square holes, as indicated.

Open the edges, so you can put your mouth between them while you hold the whistle between two fingers, and blow hard.

10.6. *Boomer*

Level of Difficulty: ★

Paper: A3/US Tabloid

For a louder boom, use the A3/US Tabloid; but if you cannot find any, it will still work with the A4/US letter.

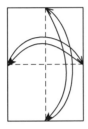 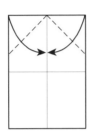 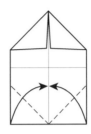 

Fold left edge to the right, and unfold. Repeat from bottom to top, and unfold.

Fold the top corners to the center line.

Repeat on the bottom.

Valley fold the top and bottom corners. This is just a mark, so there is no need to be exact.

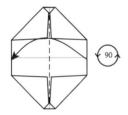 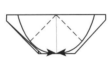

Valley fold the right edge to the left. Rotate by 90° counterclockwise.

Fold the upper edges to the center line.

Mountain fold the right edge to the left.

 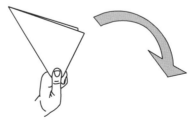

The boomer is ready!

Hold it at the missing corner, and make sure the 90° angle is pointing forward. Slash your hand forward and down in a quick and smooth movement, like hitting a table. The inner flap of the boomer will open up and make a loud noise.

10.7. Bottle Opener

Level of Difficulty: ★

Paper:

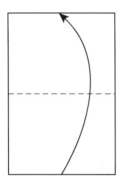
Fold bottom to top.

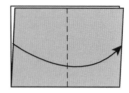
Fold left to right.

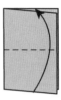
Fold bottom to top.

Fold left to right.

Fold bottom to top.

Fold **bottom to top** again.

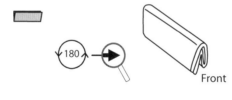

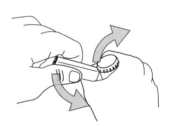

This final folded piece of paper is the bottle opener. It is almost as strong as wood.

Viewed from the side and enlarged, it is multi-layered and quite thick. Use the front end (the four-layered end) as the opener.

Hold the bottle with one hand, while tucking the front edge of the opener under the rim of the cap, and above the thumb of the holding hand. Push down the back of the opener, and let the front pop out the cap!

Chapter 10: More Fun with Paper 223

Magic Tricks

10.8. *Moving Paper*

Level of Difficulty: ★

Paper: ▢

When I (Ilan) was a child, my father showed this trick to me, and I was amazed! He 'stole' a long strand of hair from my head (although my hair was short!) and 'tied' it to the top of a strip of paper. Holding the paper in one hand, he 'pulled' the strand of hair as if it was attached to the paper, and the paper leaned toward him. He let it go, and the paper stood straight up again!

Of course, the paper moved because it was prepared beforehand. My father took a small rectangle (~2 × 5 cm) and tore it along its long axis from the top to the middle. Then, he folded one of the torn strips between his thumb and hidden finger. He slid his thumb up and down, causing the paper to bend and unbend, but I was sure there was a string of hair, I just couldn't see!

Practise this before you perform the trick!

Preparations:

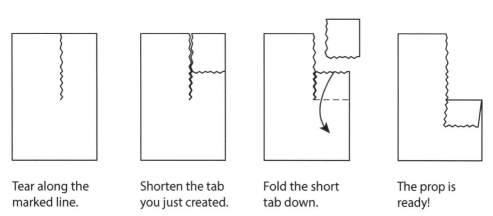

Tear along the marked line.

Shorten the tab you just created.

Fold the short tab down.

The prop is ready!

How to perform the trick:

Hold the folded tab between your thumb and finger. Make sure the tab is well hidden!

Slide your thumb down gently. The standing tab will bend inward.

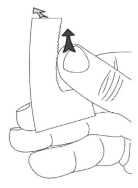

Slide your thumb back up in a sharp movement, and the standing tab will jump to an upright position.

10.9. *Joining Paperclips with a Dollar Bill*

Level of Difficulty: ★
Paper: Dollar Bill

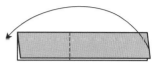

Start with a dollar bill or a sheet of paper with a ~2 : 5 proportion.

Fold the top edge to the bottom.

Valley fold the right edge to the left.

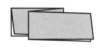
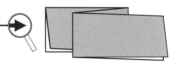

Valley fold the left edge to the right.

The bill is ready.

Enlarged view. Now add the paperclips!

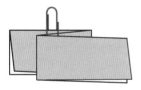
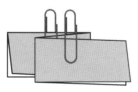
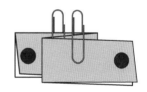

The **first** paperclip will hold the center and back layers.

The **second** should hold the front and center layers.

Now here comes the magic part! Hold the marked edges, and pull them away.

The bill is fully stretched, and look what happened to the paperclips! Without any intervention from you, they are joined! Magic!

10.10. The Upside-Down Dollar Bill

Level of Difficulty: ★
Paper: Dollar Bill

Start with Washington's head up.

Fold the top edge to the bottom.

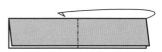
Mountain fold the right edge to the left.

Valley fold the right edge to the left.

Now, unfold the last step.

Unfold the **front** left side to the right.

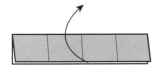
Unfold the front layer to the top.

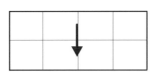
Your bill is upside-down.

And so is Washington!

The trick is hidden in the the fifth step. When you folded it, it was **backward**! Unfolding it forward makes it flip!

10.11. *Balanced Paper*

Level of Difficulty: ★★

Paper:

Balance a sheet of printer paper on your finger, straight up, without folding the paper.

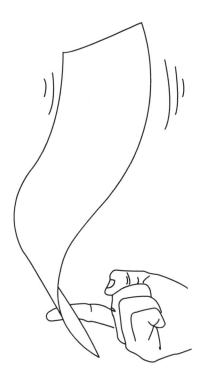

This may seem like an act of dexterity, but try and find the logic to solve this puzzle.

10.12. *A Glass on a Sheet of Paper*

Level of Difficulty: ★★

Paper:

This is a classic. You have a sheet of paper and three drinking glasses. Your task is to fill one glass with water and place it so that it is at least one 'glass-height' higher than the other two. No glass can touch any other, or to be on top of each other.

Chapter 10: Solutions

10.11. *Balanced Paper*

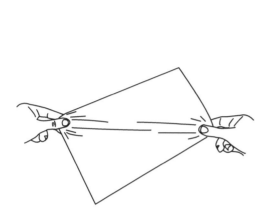
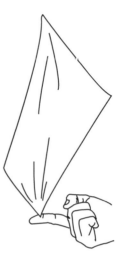

Stretch the paper from corner to corner. Try not to tear it, and create a long concave valley along the diagonal.

When the paper is distorted enough, it becomes rigid, and can be balanced on a finger, like a stick!

10.12. *A Glass on a Sheet of Paper*

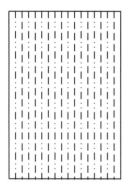
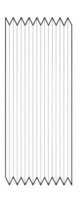
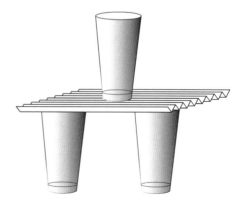

Zig-zag fold the sheet of paper, it doesn't really matter how many times.

Collapse it and put it on top of two glasses.

The new structural strength gained from the folds can carry a third glass standing on top of it.

Appendix

A.1. *Dividing a Right Angle*

Dividing into Two Equal Parts

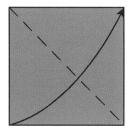

Fold corner to corner.

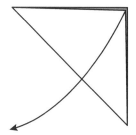

You can see the top layer exactly covers the bottom layer, meaning they are identical in size and shape. Unfold.

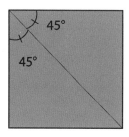

The fold line divides the right angle of the square into two equal parts.

Dividing into Three Equal Parts

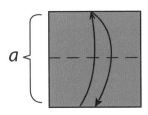

Fold edge to edge and unfold.

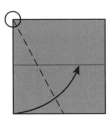

Fold bottom left corner to the crease line. Make sure the fold line starts at the top left corner!

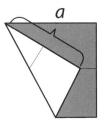

The marked edge's length is a.

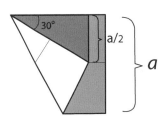

The dark gray triangle has an edge length of $a/2$ on the right.

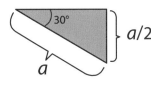

This triangle has an edge length of $a/2$ on the right and a as the longer edge. As one angle is 90°, this means the narrower angle is 30°.

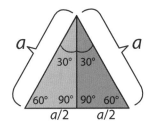

If we mirror reflect the triangle, we can see why.

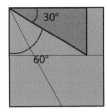

By unfolding we can see the marked angle must be 60°.

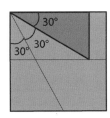

As both uncovered angles must be equal, each is 30°, so the square corner is divided into three!

A.2. *The Fujimoto Approximation — Dividing a Sheet of Paper into n Equal Parts*

Dividing a piece of paper into any number that is a power of two, is easy — fold in half, recursively. We can do this repeatedly for the paper to be divided into 2, 4, 8, 16, etc.

Dividing into even parts

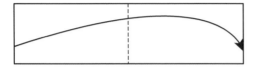 

Dividing into two is simply done by folding edge to edge.

It is easy to see that the left part totally covers the right side, and that both are equal.

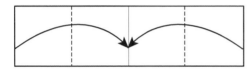 

Dividing into four parts is the same as dividing into two. Fold the edges to the center line marked by a crease.

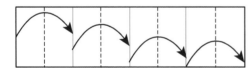 

Dividing into eight or any other power of 2 is based on the previous power.

Dividing into three parts is more complicated, and can be done by using the Origami technique. It is not an exact method, but still, after several iterations (four to six) the accuracy is more than enough.

This process was discovered by S. Fujimoto (from Japan) and it is called the **Fujimoto Approximation.**

Dividing into three parts

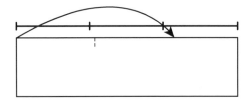 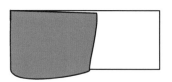

We start by trying to divide the paper into thirds, trying to make the exposed shaded side as large as the unshaded remaining portion. Do not fold all the way, only pinch the top part (marked with a short line). Since the pinch mark is only approximate, there is some error. We assume that the right (unshaded) portion is slightly larger than a third of the length of the paper by an amount we denote: Δ.

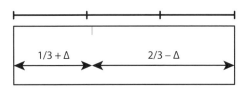

The pinch mark divides the paper into a third plus the error ($1/3 + \Delta$) on the left, and 2/3 minus the error ($2/3 - \Delta$) on the right.

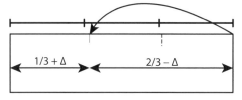

Fold the right side to the pinch mark and pinch again.

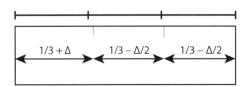

The new pinch mark divides the right side into two equal parts, each is $1/3 - \Delta/2$.

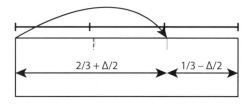

Using the new pinch mark we divide the left side now. As the right side is $1/3 - \Delta/2$, the left side must be $2/3 + \Delta/2$.

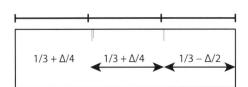

And after the third pinch, the left side is divided now into two parts, each is $1/3 + \Delta/4$!

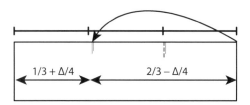

Folding the right side to the third pinch mark adds another division.

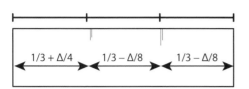 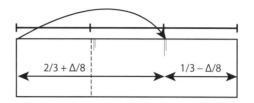

The fourth pinch mark divides the error into 8.

The fifth iteration makes the error now smaller by 16 (2^4). You can continue as you like, but at a certain point the new pinch is so close to the previous one, there is no need to continue, just complete the crease.

We can apply the same process to all odd numbers. Here is how to do this for fifths, sevenths and ninths.

Dividing into five parts

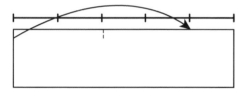 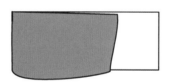

To divide by five, your first pinch is trying to divide the paper into 2/5 and 3/5 (make the white side approximately half the shaded side).

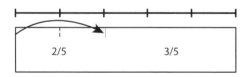 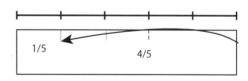

2/5 can be easily divided into 1/5.

Marking the 1/5 on the left, leaves us 4/5 on the right, which can be just as easily divided into 2/5.

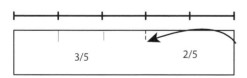

And the 2/5 is divided into 1/5. Repeat the process until the pinch marks are all aligned.

Dividing into seven parts

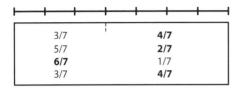

3/7	**4/7**
5/7	**2/7**
6/7	1/7
3/7	**4/7**

Dividing into nine parts

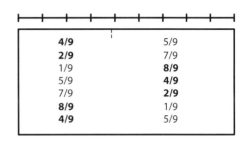

4/9	5/9
2/9	7/9
1/9	**8/9**
5/9	**4/9**
7/9	**2/9**
8/9	1/9
4/9	5/9

To divide by seven, your first pinch mark is trying to divide the paper into 3/7 and 4/7. The next steps are to fold to half the side with the even number of sevenths.

To divide by nine, your first pinch mark is trying to divide the paper into 4/9 and 5/9. The next steps are to fold to half the side with the even number of ninths.

A.3. The 11 Nets of a Cube

References

Books and Articles

1. British Origami Society (BOS) Magazines.
2. E. D. Demaine and J. O'Rourke, *Geometric Folding Algorithms* (Cambridge University Press, New York, 2007).
3. M. Driscoll, *Fostering Geometric Thinking: A Guide for Teachers, Grades 5–10* (Heinemann, Portsmouth, 2007).
4. H. E. Dudeney, *536 Puzzles and Curious Problems* (Dover Publications, New York, 2016) (First published by Charles Scribner's Sons, New York, 1967).
5. K. Fulves, *Self-Working Paper Magic: 81 Fullproof Tricks* (Dover Publications, New York, 1985).
6. M. Gardner, *Wheels, Life and Other Mathematical Amusements* (W. H. Freeman and Company, San Francisco, 1983).
7. M. Gardner, *Martin Gardner's New Mathematical Diversions from "Scientific American"* (Allen & Unwin, London, 1969).
8. M. Gardner, *Mathematical Magic Show: More Puzzles, Games, Diversions, Illusions & Other Mathematical Sleight-of-Mind from Scientific American*, 1st Vintage Books edn. (Vintage Books, New York, 1978).
9. M. Gardner, *Hexaflexagons and Other Mathematical Diversions: The First Scientific American Book of Puzzles & Games* (University of Chicago Press, Chicago, 1988).
10. M. Gardner, *Entertaining Science Experiments with Everyday Objects* (Dover Publications, New York, 1981).
11. P. J. Hilton, J. Pedersen, and S. Donmoyer, *A Mathematical Tapestry: Demonstrating the Beautiful Unity of Mathematics* (Cambridge University Press, New York, 2010).
12. D. Mitchell, *Paperfolding Puzzles* (Water Trade Publications, 2011).

13. J. Montroll, *Origami & Geometry: Stars, Boxes, Troublewits, Chess, & More* (Antroll Publishing Company in association with CreateSpace Independent Publishing Platform, 2013).
14. R. E. Neale, Self-designing tetraflexagon, in *The Mathematician and Pied Puzzler: A Collection in Tribute to Martin Gardner*, eds. E. R. Berlekamp, and T. Rogers (A. K. Peters, Natick, MA, 1999), pp. 117–126.
15. C. A. Pickover, *The Möbius Strip* (Thunder's Mouth Press, New York, 2006).
16. L. Pook, *Serious Fun with Flexagons* (Springer, New York, 2009).

Internet Sites

1. The Unique Projects website: http://www.uniqueprojects.com/
2. http://www.cutoutfoldup.com/index.php
3. The Strip to Square puzzle:
   http://www.archimedes-lab.org/monthly_puzzles_62.html#
4. http://cutnfold.net/